Fish Stock Assessment Manual

by
Emygdio L. Cadima
Consultant
FAO Fisheries Department

Reprinted by arrangement with the
Food and Agriculture Organization of the United Nations

by
Daya Publishing House®
A Division of
Astral International Pvt. Ltd.
New Delhi – 110 002

"This book was originally published by the Food and Agriculture Organization of the United Nations (FAO) as FAO Fisheries Technical Paper No., 393".

First Indian Edition, 2007
ISBN: 9788170354963

Published by : **Daya Publishing House®**
A Division of
Astral International Pvt. Ltd.
– ISO 9001:2015 Certified Company –
4736/23, Ansari Road, Darya Ganj
New Delhi-110 002
Ph. 011-43549197, 23278134
E-mail: info@astralint.com
Website: www.astralint.com

Digitally Printed at : **Replika Press Pvt. Ltd.**

PREPARATION OF THIS DOCUMENT

The author, Emygdio Cadima, now retired, was an FAO scientist in the Fisheries Department until 1974, when he returned to the Instituto de Investigação das Pescas e do Mar (IPIMAR) in Portugal, having also been a Professor at the University of Algarve until 1997. At the end of 1997 he was the lecturer of a course in Fish Stock Assessment in IPIMAR, which became the basis for the preparation of this manual, requested and supported by the Project FAO/DANIDA GCP/INT/575/DEN. This manual also incorporates notes from courses in Fish Stock Assessment held at several different venues in the world, mainly in Europe, Latin America and Africa. These courses had an active collaboration of fisheries scientists from all over the world, especially Portugal. These scientists are also co-responsible for the orientation, for the matters treated and particularly for the elaboration of the exercises.

This manual aims to present the basic knowledge on the problems and methods of fish stock assessment to young scientists, post-graduate students, and PhD students. This is a scientific area in permanent development, where the knowledge of fisheries biology is applied in order to make a rational and sustained exploitation of the fishing resources.

The "Manual of Fish Stock Assessment" is mainly concerned with the theoretical aspects of the most used models for fish stock assessment. The practical application (i.e. the exercises solved in a spreadsheet), is considered as a complementary part to help the understanding of the theoretical matters.

The editing of the manuscript was made by Siebren Venema, manager of Project GCP/INT/575/DEN and Ana Maria Caramelo, Fishery Resources Officer in the FAO Fisheries Department.

Distribution:

DANIDA
Fisheries Education Institutes
Marine Research Institutes
National and International Organizations
Universities
FAO Fisheries Department

Cadima, E.L.
Fish stock assessment manual.
FAO Fisheries Technical Paper. No. 393. Rome, FAO. 2003. 161p.

ABSTRACT

The manual follows the same order of the lectures in the last course held in IPIMAR (November/December 1997). It starts with an introduction to the mathematical models applied in Fish Stock Assessment and some considerations on the importance of fisheries. The need for a rational management of the fishing resources is then stressed, this being indispensable for an adequate exploitation, aiming at conservation, to occur. The basic assumptions about a model and the concepts of different variation rates of a characteristic in relation to time (or to other characteristics) are presented, highlighting the most important aspects of the simple and exponential linear models which are used in the chapters that follow. After some considerations on the concept of cohort, models for the evolution in time of the number and weight of the individuals that constitute the cohort are developed, including models for the individual growth of the cohort. In the chapter concerning the study of the stock, the fishing pattern and its components are defined, the most used models for the stock–recruitment relation are presented, as well as the short and long term projections of a stock. With regard to fishing resources management, the discussion is focused on the biological reference points (target points, limit points and precautionary points) and fisheries regulation measures. The last chapter, which presents and discusses theoretical models of fish stock assessment, deals with production models (also designated as general production models) and with the long and short–term projections of the catches and biomasses. Finally, the general methods of estimating parameters are described and some of the most important methods are presented, with special relevance to the cohort analysis by age and length. Then a solution of the exercises from the last course held in IPIMAR, is presented by the author and the scientist Manuela Azevedo.

TO MY FIRST MASTERS AND OLD-TIME FRIENDS

Ray Beverton
John Gulland
Gunnar Sætersdal

PREFACE

This work is essentially orientated to present an introduction to the mathematical models applied to fisheries stock assessment.

There are several types of courses about the methods used in fish stock assessment.

One type considers practical application as the main aspect of the course, including the use of computer programs. The theoretical aspects are referred to and treated as complementary aspects.

A second type is mainly concerned with the theoretical aspects of the most used models. The practical application, considered as the complementary part, facilitates the understanding of the theoretical subjects.

In this work, the second type was adopted and exercises were prepared to be solved in a worksheet (Microsoft Excel). The table of contents indicates the exercises corresponding to each subject.

This manual is the result of a series of courses on Fish Stock Assessment held in the following places. Portugal : Instituto de Investigação das Pescas e do Mar – IPIMAR (ex-INIP) in Lisbon, Faculdade de Ciências de Lisboa, University of Algarve and Instituto de Ciências Biomédicas de Abel Salazar in Oporto. Other courses were held at Instituto de Investigação das Pescas in Cape Verde, at the Centro de Investigação Pesqueira in Angola, at the Instituto de Investigação das Pescas in Mozambique, at the Centro de Investigacion Pesquera – CIP in Cuba, at the Instituto del Mar del Perú – IMARPE in Peru, at the Instituto Español de Oceanografía – IEO (Vigo and Málaga – Spain). It is also a result of some lectures integrated into cooperation courses held in several countries and organized by FAO, by SIDA (Sweden), by NORAD (Norway) and by ICCAT.

Other fisheries scientists cooperated in these courses and they are also co-responsible for the orientation of the subjects studied and very particularly for the elaboration of the exercises and the editorial work. With no particular criterium, these are some of the collaborators to whom I express my appreciation: Ana Maria Caramelo, Manuel Afonso Dias, Pedro Conte de Barros, Manuela Azevedo Lebre, Raúl Coyula, Renato Guevara.

Lisbon, December 1997
E. Cadima

CONTENTS

CONTENTS

CONTENTS

GLOSSARY OF TECHNICAL TERMS USED IN THE MANUAL

Abundance index (U) – A characteristic preferably proportional to the available biomass of the resource. The catch per unit effort, **cpue** (especially when the effort is expressed in appropriate units) is an important index..

Biological Limit Reference Point (LRP) – Biological reference point indicating *limits* of the fishery exploitation with regard to stock self-reproduction, aiming at conservation of the resource.

Biological Precautionary Reference Point (PaRP) – biomass levels (**Bpa**) and fishing levels (**Fpa**), established under the precautionary principle, concerning the reproduction of the stock, aiming at conservation of the resources. The assumptions and methods used to determine the PaRPs should be mentioned.

Biological Reference Point (BRP) – Values of **F** and **B**, taking into consideration the *best possible catch* and/or ensuring the *conservation* of the fishery resource. There are BRPs based on long term projections (LP), BRPs based on values observed during a certain period of years and BRPs based on the two previous criteria. The BRPs can be *Target-Points (TRP), Limit-Points (LRP), and Precautionary Points (PaRP)*. In this manual the following biological reference points are referred to: F_{max}, $F_{0.1}$, F_{high}, F_{med}, F_{MSY}, F_{loss}, F_{crash}, B_{max}, $B_{0.1}$, B_{med}, B_{MSY}, B_{loss}, **MBAL.** Other biological reference points, used in management, like $F_{30\%SPR}$, are not mentioned in this manual.

Biological Target Reference Point (TRP) – Biological reference point indicating long term o*bjectives* (or targets), for the management of a fishery, taking into consideration the *best possible catch* and ensuring the conservation of the stock.

Biomass (B) – Weight of an individual or a group of individuals contemporaneous of a stock.

Capturability Coefficient (q) – Fraction of the biomass that is caught by unit of fishing effort.

Carrying capacity (k) – Capacity of the environment to maintain the stock living in it. It is, theoretically, the limit of the non exploited biomass (see intrinsic gross rate of the biomass, **r**).

Catch in number (C) – Number of individuals caught.

Catch in weight or **Yield (Y)** – Biomass of the stock taken by fishing. Yield does not necessarily correspond to *landed weight*. The difference between the two values, yield and landings, is mainly due to *rejections to the sea* of part of the catch which, for some reason (price, quality, space problems or even legal reasons), is not landed.

Cohort – Set of individuals of a fishery resource born from the same spawning.

Exploitation pattern of a gear (s) – Fraction of the individuals of a given size, available to the gear, which is caught. Also designated by *Selectivity* or *partial recruitment.*

Individual growth coefficient (K) – Instantaneous relative rate of change of a function of the individual weight, w, that is, **H(w$_\infty$)-H(w),** where **w$_\infty$** is the *asymptotic individual weight* and **H(w)** is a function of **w** (frequently a power function, including the logarithmic function). The adopted models for the function **H(w)** have two constants, **w$_\infty$** and **K**. Some models introduce one more parameter, **b**, which is used to obtain a general relation to include the most common individual growth relations. The constant **K** has the physical dimension of time $^{-1}$.

Individual Quota (IQ) – Quota attributed to a vessel.

Individual Transferable Quotas (ITQ) – System of fisheries management characterized by the sale, at auctions, of the fishing annual vessel quotas.

Minimum Biomass Acceptable Level (MBAL) – Biological reference limit point that indicates a spawning biomass level under which the observed biomasses during a period of years, are small and the associated recruitments are smaller than the mean or median recruitment.

Number of individuals of a cohort or of a stock (N) – Number of survivals of a cohort (or a stock) at a certain instant or over an interval of time.

Partial recruitment – (see exploitation pattern)

Precautionary principle – This principle establishes that a lack of information does not justify the absence of management measures. On the contrary, management measures should be established in order to maintain the conservation of the resources. The assumptions and methods used for the determination of the scientific basis of the management should be presented.

Production models – Models that consider the biomass of the stock as a whole, that is, they do not take into consideration the age or size structure of the stock. These models are only applied in analyses that consider fishing level changes, as they do not allow the analysis of the effects of changes in the exploitation pattern, on catches and biomasses.

Quota (Q) – Each of the fractions in which the TAC was divided.

RATES

Absolute Instantaneous Rate of y, air(y) – Velocity of the variation of the function **y(x)**, at the instant **x**.

Absolute Mean Rate of y, amr(y) – Mean velocity of the variation of the function **y(x),** during a certain interval of **x**.

Annual Survival Rate (S) – Mean rate of survivals of a cohort during one year, relative to the initial number.

Exploitation Rate (E) – Ratio between the number of individuals caught and the total number of individuals dead, over a certain period of time, that is, **E = C/D**.

Fishing mortality instantaneous rate (F) (Fishing mortality coefficient) – Relative instantaneous rate of the mortality of the number of individuals that die due to fishing .

Intrinsic rate of the biomass growth (r) – Constant of the *Production models* that represents the instantaneous rate of the decreasing of the function **H(K)-H(B)**, where **B** is the biomass, **H(B)** is a function of the total biomass, usually a power-function, (including the logarithmic function that can be considered a limit power function) and **k** is the **carrying capacity** of the environment. Some models introduce one more parameter, **p**, which is used to obtain a more general relation.

Natural mortality instantaneous rate (M) (Natural mortality Coefficient) – Instantaneous relative rate of the mortality of the number of individuals that die due to all causes other than fishing.

Relative instantaneous rate of y, rir(y) – Velocity of the variation of the function **y(x),** relative to the value of **y,** at the instant **x**.

Relative mean rate of y, rmr(y) – Mean velocity of the variation of the function **y(x)** relative to a value of y, during a certain interval of **x**.

Total mortality instantaneous rate (Z) (Total mortality coefficient) – Relative instantaneous rate of the mortality of the number of individuals that die due to all causes. **Z, F** and **M** are related by the following expression : **Z=F+M** .

Recruitment to the exploitable phase (R) – Number of individuals of a stock that enter the fishery area for the first time each year.

Selectivity – (see exploitation pattern)

Spawning or **adult biomassa (SP)** – Biomasss of the stock (or of a cohort) which has already spawned at least once.

Stock – Set of survivals of the cohorts of a fishery resource, at a certain instant or period of time. It may concern the biomass or the number of individuals.

Stock-Recruitment (S-R) relation – Relation between the parental stock (spawning biomass) and the resulting recruitment (usually the number of recruits to the exploitable phase). The models have two constants, α and **k**. The constant **k** has the physical dimension of weight and α has the dimension of weight^{-1}. Some models introduce one more parameter, **c**, which is used to obtain a general relation that includes the most common relations.

Structural models – Models that consider the structure of the stock by ages or sizes. These models allow one to analyse the effects on catches and biomasses, due to changes in the fishing level and exploitation pattern.

Total Allowable Catch (TAC) – Management measure that limits the total annual catch of a fishery resource, aiming to indirectly limit the fishing mortality. The TAC can be divided into *Quotas (Q)* using different criteria, like countries, regions, fleets or vessels.

Total number of deaths (D) – Total number of individuals that die during a certain period of time..

Virgin biomass (VB) – Biomass of the stock not yet exploited.

SYMBOLS

Symbols	Indicating :
A	Constant of the simple linear model (intercept of the straight line)
α	Constant of the Stock-Recruitment relations (limit value of R/S when $S \to 0$)
amr(y)	Absolute mean rate of variation of y
air(y)	Absolute instantaneous rate of variation of y
B	Constant of the simple linear model (slope of the straight line)
B	Biomass
SP, SB	Spawning Biomass
C	Catch, in number
C	Constant of the Stock-Recruitment relations (generalizes the models)
D	Total number of deaths
E	Exploitation rate
F	Fishing mortality coefficient
Cconst	Non defined constant
Cte	Non defined constant
H	General power function
ITQ	Individual Transferable Quotas
K	Constant of the individual growth models (associated to growth rate)
k	Constant of the Stock-Recruitment relations
k	Constant of the production models (Carrying capacity)
L,L	Total length of an individual
MBAL	Minimum Biomass Acceptable Level (biological reference limit point)
M	Natural mortality coefficient
N	Number of individuals of a cohort
P	Constant of the Production models (generalizes the models)
Q	Capturability coefficient
R	Constant of the Production models (intrinsic rate associated with the biomass growth)
R^2	Determination coefficient
R	Recruitment to the exploitable phase
rmr(y)	Relative mean rate of variation of y
rir(y)	Relative instantaneous rate of variation of y

Symbols	Indicating :
S	Annual Survival rate
S	Adult or total biomass (in the relations S-R)
s	Exploitation pattern (selectivity)
SQ	Sum of the squares of the deviations
S-R	Stock-Recruitment relation
t	Instant of time
T	Interval of time between 2 instants
TAC	Total Allowed Catch
TRP	Biological Target Reference Point
U	Stock abundance index
V	Function to be maximized for the determination of $F_{0.1}$
W	Individual weight
Y	Catch in weight
Z	Total mortality coefficient (total mortality instantaneous rate)

SUBSCRIPTS

The characteristics of this glossary are usually shown with indices; that is why it was considered necessary to present the meaning of those subscripts.

Subscripts	**Indicating :**
$	Economical value of the respective characteristic of the cohort
λ	Maximum age
0.1	Value of F (and of other characteristics of the cohort) corresponding to the air of the biomass equal to 10 percent of the virgin biomass
c	Recruitment to exploitable phase
crash	Value of F which, at long term, corresponds to the collapse value of the spawning biomass
E	Value of the characteristics of the cohort corresponding to an equilibrium point
i	Age
infl	Value of the characteristic corresponding to an inflection point of any relation between that characteristic and other variable.
l	Length
lim	Value of B or of F corresponding to a biological reference limit point
loss	Value of B or of F corresponding to the minimum spawning biomass observed
Max	Value of F (and of other characteristics of the cohort) where the yield per recruit is maximum
Med	Value of F (and of other characteristics of the cohort) which, at long term, will produce a spawning biomass per recruit equal to the median value of the spawning biomasses per recruit observed during a certain period of years
MSY	Value of F (and of other characteristics of the stock) where the long term total yield is maximum
R	Recruitment to the exploitable phase

BIBLIOGRAPHY

Bertalanffy, L. Von 1938. A quantitative theory of organic growth. *Hum. Biol.*, 10 (2): 181-213.

Beverton, R.J.H. & Holt, S.J. 1956. On the dynamics of exploited fish populations. U.K. Min. Agric. Fish., *Fish. Invest* (Ser. 2) 19: 533p.

Caddy, J.F.& Mahon, R. 1995. Reference points for fishery management. *FAO Fish. Tech. Pap.* 349: 80p.

Cadima, E. 1991. Some relationships among biological reference points in general production models. *ICCAT, Coll. Vol. Sc. Papers*, (39):27-30.

Cadima, E. & Pinho, M.R. 1995. Some theoretical considerations on non equilibrium production models. *ICCAT, Coll. Vol. Sc. Papers*, (45):377-384.

Cadima, E. & Palma, C. 1997. Cohort analysis from annual length catch compositions. WD presented to the Working Group on the assessment of the Southern Shelf Demersal Stocks. Copenhagen, 1-10 September, 1997.

Cadima, E. & Azevedo, M. 1998. A proposal to select reference points for long term fishery management objectives. ICES,C.M. 1998/T:9, 18p.

Cardador, F. 1988. Estratégias de exploração do stock de pescada, *Merluccius merluccius* L., das águas Ibero-Atlânticas. Efeitos em stocks associados. Dissertação apresentada para provas de acesso à categoria de Investigador Auxiliar. IPIMAR (*mimeo*)

CE 1994. Report of the southern hake task force. Lisbon, 10-14 October, 1994.

Clark, W.G. 1991. Groundfish exploitation rates based on life history parameters. *Can. J. Fish. Aquat. Sci.*, 48: 734-750.

Clark, W.G. 1993. The effect of recruitment variability on the choice of target level of spawning biomass per recruit. pp 233-246 *In:* Kruse, G.; Eggers, D.M.; Marasco, R.J.; Pautzke, C. & Quinn, T.J. (eds.). Proceedings of the International Symposium on Management Stategies for Exploited Fish Populations. *Alaska Sea Grant College Program Report* Nº 93-02, Fairbanks, University of Alaska.

Cushing, D.H. 1996. Towards a science of recruitment in fish populations. *In:* Excelence in Ecology, Book 7, Ecology Institut, Oldendorf/Scuhe, Germany.

Deriso, R.B. 1980. Harvesting strategies and parameter estimation for an age-structured model. *Can. J. Fish. Aquat. Sci.*, 37: 268-282.

Duarte, R; Azevedo, M. & Pereda, P. 1997. Study of the growth of southern black and white monkfish stocks. *ICES J. mar. Sci.*, 54: 866-874.

FAO 1995. Code of Conduct for Responsible Fisheries, Rome, FAO, 41p.

FAO 1996. Precautionary approach to fisheries. *FAO Fish. Tech. Pap.* 350 (2): 210p.

Ford, E. 1933. An account of the herring investigations conducted at Plymouth during the years from 1924-1933. *J. Mar. Biol. Assoc.*, N.S., 19: 305-384.

Fox, W.W. Jr 1970. An exponential surplus-yield model for optimizing exploited fish populations. *Trans. Am. Fish Soc.*, 99: 80-88.

Garrod, G.J. 1969. Empirical assessment of catch/effort relationships in the North Atlantic cod stocks. *Res. Bull ICNAF*, 6: 26-34.

Gompertz, B. 1825. On the nature of the function expressive of the law of human mortality, and on a new mode of determining the value of life contingencies. *Phil. Trans. Royal Society*, 115 (1): 513-585.

Gulland, J.A.1959. Fish Stock Assessment: A manual of basic methods. FAO/Wiley series, 223p.

Gulland, J.A. 1969. Manual of Methods for Fish Stock Assessment - Part 1. Fish Population Analysis. *FAO Manuals in Fisheries Science No. 4.*

Gulland, J.A. 1983. *Fish stock assessment A manual of basic methods.* FAO/Wiley Ser. on Food and Agriculture, Vol 1: 233 pp.

Gulland, J.A. & Boerema, L.K 1973. Scientific advice on catch levels. *Fish.Bull.* 71 (2): 325-335

Gulland, J.A. & Holt, S.J. 1959. Estimation of growth parameters for data at unequal time intervals. *J. Cons. ICES*, 25 (1): 47-49.

Gunderson, D.R. 1980. Using r-K selection theory to predict natural mortality. *Can. J. Fish Aquat. Sci.* 37: 2266-2271.

Hilborn, R. & Walters, C.J. 1992. *Quantitative Fisheries Stock Assessment: Choice, Dynamics and Uncertainty*. New York, Chapman and Hall, 570p.

ICES 1996. Report of the Northern Pelagic and Blue Whiting Fisheries Working Group. Bergen, 23-29 April 1996. ICES CM 1996/Assess:14

ICES 1997a. Report of the Working Group on the Assessment of Southern Shelf Demersal Stocks. Copenhagen, 3-12 September 1996. ICES CM 1997/ASSESS:5

ICES 1997b. Report of the Working Group on the Assessment of Mackerel, Horse Mackerel, Sardine and Anchovy. Copenhagen, 13-22 August 1996. ICES CM 1997/ASSESS:3

ICES 1997c. Report of the Working Group on the Assessment of Southern Shelf Demersal Stocks. Copenhagen, 3-12 September 1996. ICES CM 1997/ASSESS:5

ICES 1997d. Report of the Comprehensive Fishery Evaluation Working Group. Copenhagen, 25 June - 04 July 1997. ICES CM 1997/ASSESS:15.

ICES 1998a. Report of the Study Group on the Precautionary Approach tro Fisheries Management. Copenhagen, 3-6 February 1998. ICES CM 1998/ACFM: 10

ICES 1998b. Report of the Working Group on the Assessment of Mackerel, Horse Mackerel, Sardine and Anchovy. Copenhagen, 9-18 September 1997. ICES CM 1998/ASSESS:6

ICES 1998c. Report of the Working Group on the Assessment of Southern Shelf Demersal Stocks. Copenhagen, 1-10 September 1997. ICES CM 1998/ASSESS:4

ICES 1998d. Report of the Working Group on the Assessment of Northern Shelf Demersal Stocks. Copenhagen, 16-25 June 1997. ICES CM 1998/ASSESS:1

Jones, R. 1961. The assessment of the long term effects of changes in gear selectivity and fishing effort. *Mar. Res. Scot.*, 2, 19p.

Jones, R. & van Zalinge, N.P. 1981. Estimates of mortality rate and population size for shrimp in Kuwait waters. *Kuwait Bull. Mar. Sci.*, 2: 273-288.

Lotka, A.J. 1925. *The Elements of Physical Biology*. Baltimore, Williams and Wilkins.

Marquardt, D.W. 1963. An algorithm for least squares estimation of non-linear parameters. *J. Soc. Ind. Appl. Math.*, 2: 431-441.

Mattos e Silva, G.O. 1995. Aplicação de modelos de produção geral em condições de não-equilíbrio para a avaliação do manancial de gamba *Parapenaeus longirostris* (Lucas, 1846) da costa sul portuguesa. Dissertação apresentada para obtenção do grau de Mestre em Estudos Marinhos e Costeiros. UAL, Unidade de Ciências e Tecnologias dos Recursos Aquáticos, Faro, 96p.

Megrey, B. 1989. Review and comparison of age-structured stock assessment models from theoretical and applied points of view. *Am. Fish. Soc. Symp.*, 6: 8-48.

Paloheimo, J.E. 1961. Studies on estimation of mortalities. Comparison of a method described by Beverton and Holt and a new linear formula. *J. Fish. Res. Bd. Can.*, 18 (5): 645-662.

Pauly, D. 1980. On the interrelationships between natural mortality, growth parameters and mean environmental temperature in 175 fish stocks. *J. Cons. Int. Explor. Mer*, 39: 175-192.

Pella, J.J. & Tomlinson, P.K. 1969. A generalized stock production model. *Bull. Inter. Am. Trop. Tuna Comm.*, 13: 419-496.

Pestana, G. 1989. Manancial Ibero-Atlântico de sardinha, *Sardina pilchardus*, Walb., sua avaliação e medidas de gestão. Dissertação original para provas de acesso à categoria de Investigador Auxiliar. IPIMAR, 192p. (*mimeo*).

Pope, J.G. 1972. An investigation of the accuracy of virtual population analysis using cohort analysis. *Res. Bull. ICNAF*, 9: 65-74.

Prager, M.H. 1994. A nonequilibrium surplus-production model. *Fish. Bull.* 92 (2): 372-389

Prager, M.H. 1995. User´s Manual for ASPIC: a stock-production model incorporating covariates, program version 3.6x. Miami Lab. Doc. MIA-92/93-55

Richards, F.J. 1959. A flexible growth function for empirical use. *J. Exp. Bot.*, 10: 290-300.

Ricker, W.E. 1954. Stock and recruitment. *J. Fish. Res. Bd. Can.*, 11: 559-623.

Ricker, W.E. 1958. Handbook of computation for biological statistical of fish population. *Bull. Fish. Res. Bd. Can.*, 119: 300p.

Ricker, W.E. 1969. Effects of size-selective mortality and sampling bias on estimates of growth, mortality, production and yield. *J. Fish. Res. Bd. Can.*, 26: 479-541.

Ricker, W.E. 1975. Computation and interpretation of biological statistics of fish population. *Bull. Fish. Res. Bd. Can.*, 191: 382p

Rikhter, J.A. & Efanov, V.N. 1976. On one of the approaches to estimation of natural mortality of fish population. *ICNAF* 76/VI/8, 12p.

Rosenberg, A.A.; Kirkwood, G.P.; Crombie, J.A. & Beddington, J.P. 1990. The assessment of stocks of annual squid species. *Fish. Res.* 18:335-350.

Sætersdal, G. 1984. Investigação, gestão e planificação pesqueiras. *Revista de Investigação Pesqueira*, 9. Instituto de Investigação Pesqueira. Maputo. R.P.M.: 167-186.

Schaefer, M. 1954. Some aspects of the dynamics of populations important to the management of the commercial marine fisheries. *Bull. Inter. Am. Trop. Tuna Comm.*, 1 (2): 27-56.

Shepherd, J.G. 1982. A versatile new stock-recruitment relationship for fisheries, and the construction of sustainable yield curves. *J. Cons. Int. Explor. Mer*, 40 (1): 67-75.

Sparre, P. & Venema, S.C. 1997. Introdução à avaliação de mananciais de peixes tropicais. *FAO Doc. Téc.Pescas*, 306/1 Rev 2. (Parte 1 & 2): 404 & 94 pp.

Stamatopoulos, C. & Caddy, J.F. 1989. Estimation of Von Bertalanffy growth parameters: a versatile linear regression approach. *J. Cons. Int. Explor. Mer*, 45: 200-208.

Tanaka, S. 1960. Studies on the dynamics and the management of fish populations. *Bull. Tokai. Reg. Fish. Res. Lab.*, 28: 1-200.

Volterra, V. 1928. Variations and fluctuations of the number of individuals in animal species living together. *J. Cons. Int. Expl. Mer*, 3 (1): 3-51.

Yoshimoto, S.S. & Clarke, R.P. 1993. Comparing dynamic versions of the Schaefer and Fox production models and their application to lobster fisheries. *Can. J. Fish. Aquat. Sci.* 50: 181-189.

Walford, L.A. 1946. A new graphic method of describing the growth of animals. *Biol. Bull. Mar. Biol. Lab. Woods Hole*, 90: 141-147.

CHAPTER 1 – INTRODUCTION

1.1 THE IMPORTANCE OF FISHERIES

The importance of fisheries in a country cannot only be measured by the contribution to the GDP, but one must also take into consideration that fisheries resources and products are fundamental components of human feeding and employment.

Another aspect that makes fisheries resources important is the self renewable character. Unlike mineral resources, if the fishery resources or any other biological resources are well managed, their duration is pratically unlimited.

An important conclusion is that the fundamental basis for the conservation and management of fisheries resources stems from the biological characteristics. (This does not mean that social, economic or any other effects are not important for management).

In Portugal, the fisheries contribution to the GDP is less than 1.5 percent. However, with regard to food, the annual consumption value of 60 kg of fish per person, is very high. Only countries like Iceland, Japan and some small insular nations reach a higher value. We still have to consider that of the total amount of protein necessary in our food consumption, 40 percent comes from fisheries. This corresponds to 15 percent of the total amount spent on food by the Portuguese population.

From a social point of view, we estimate that there are, at present, 34 000 fishermen in Portugal. Assuming that each job at sea generates 4 or 5 jobs on land (canning, freezing and fish meal industry, commercialization, administration, research and training, etc.) one can estimate that about 150 000 Portuguese work in the several sectors of fisheries. Consequently, taking a minimum of 3 people per family, it is not unreasonable to say that about half a million Portuguese people depend on fisheries activities for their livelihoods.

1.2 FISHERIES RESOURCES MANAGEMENT

Sætersdal (1984) defined a general principle of fisheries management as :

"to obtain the BEST POSSIBLE utilization of the resource for the benefit of the COMMUNITY"

It will be necessary to define, in each particular case, what best, possible and community mean.

In fact, best can be taken as :

- Bigger yield
- Bigger value of the catch
- Bigger profit (difference between the value of the landing and the costs of exploitation)
- More foreign currency
- More jobs, etc.

Community may also be taken as :

- The population of the world
- The European Community
- A country
- A region
- Groups of interests (fishermen, shipowners, consumers, …)

Possible

reminds us that we cannot forget the self renewable character of fisheries resources and consequently, that the conservation of the fishery resource must be guaranteed in order to allow the application of the general principle for a long period of time. This statement means that conservation of an ecosystem does not imply that one should attribute the same importance to all its components.

1.3 FISHERIES RESOURCES RESEARCH

Figure 1.1 shows that the research on fisheries resources covers several sectors of the fishing activity.The assessment models are the main concern of this manual. Among the several works and books on fish stock assessment, the books and/or manuals by Beverton & Holt (1956), Ricker (1958, 1975) and Gulland (1969, 1983) are historical standing references.

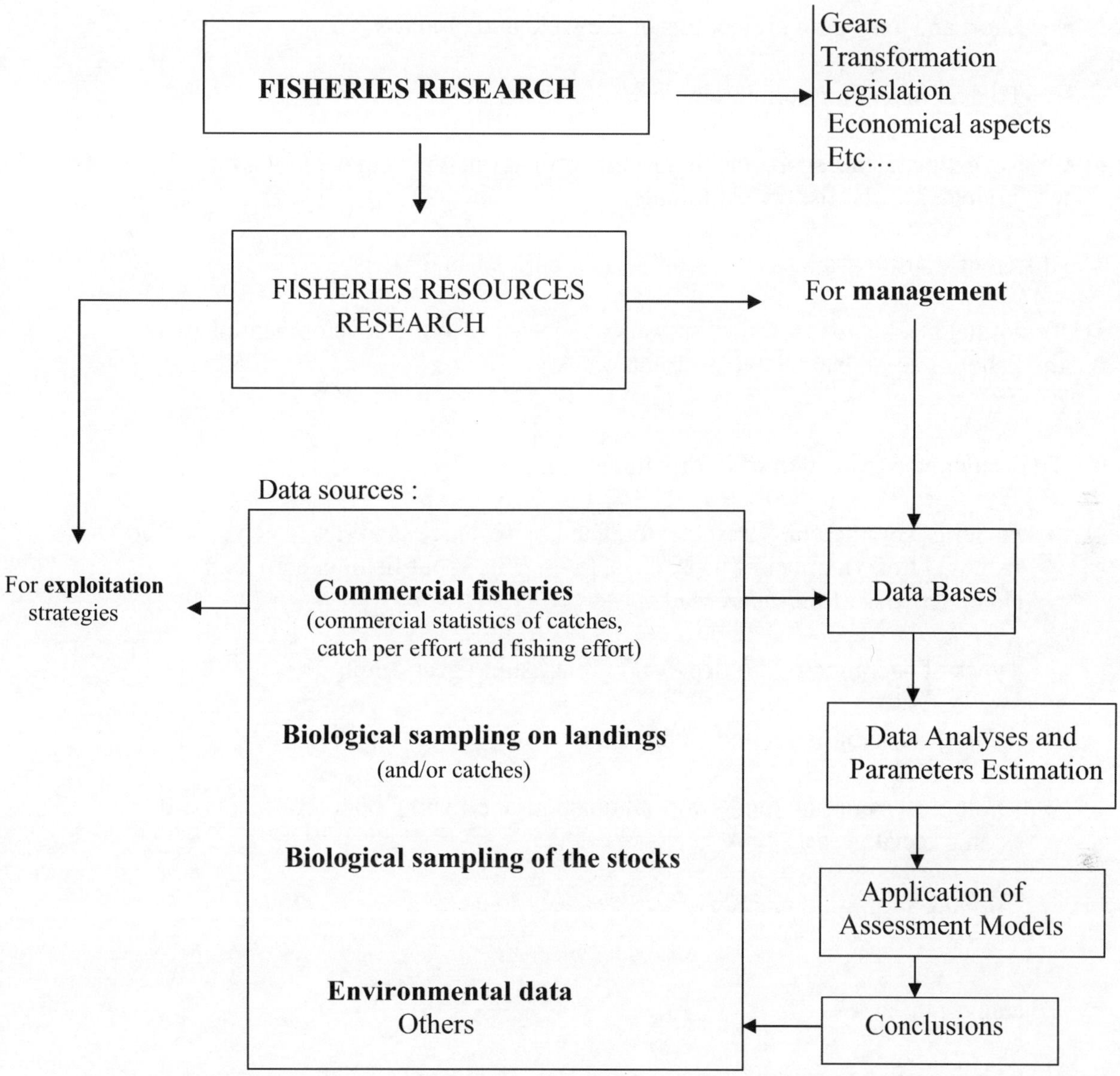

Figure 1.1 The several sectors of fishing activity

1.4 FISH STOCK ASSESSMENT

The following are necessary to assess a fish stock:

- The appropriate data bases
- Analyses of the available data

- Short and long-term projections of the yield and biomass
- To determine long-term biological reference points
- To estimate the short and long-term effects on yield and biomass of different strategies of the fishery exploitation

The different steps to assess a stock can be summarized as follows :

a) To define the objectives of the assessment according to the development phase of the fisheries and the available information.

b) To promote the collection of information :

- Fisheries commercial statistics : total and by resource landings, catch per effort, fishing effort (number of trips, days, tows, time spent fishing, etc.), and characteristics of the gears used.
- Types of operation of the fleets and of its fishing gears, etc.
- Biological sampling in the landing ports.
- Biological sampling (and information about the fishing operation) on board commercial vessels.
- Biological sampling on board research vessels.

c) To analyse the stocks

The knowledge gained about the resource and the available basic data, determine the type of models that should be used and consequently the type of analyses that can be done. As an illustration, let us look at some general situations :

Fishery resource with little information

↓

Analyses using particular methods to estimate biomasses and potential yields.

Fishery resource with data on catches and catch per effort (CPUE) or stock abundance indices during several years

↓

Analyses using production models in order to make projections of yield and catch per effort.

Fishery resource with information collected over several years on :

- Biological distribution of the catches by species, by length, by ages, etc.
- Commercial catches
- Fishing effort or CPUE
- Research cruises (distribution of the stock by areas, by length, by ages, etc.)

↓

Historical analysis of the stock (VPA)
Long and short-term projections with different conditions (scenarios)

Comments

1. The lack of information may prevent certain projections, but allows other types of analyses.

2. The Precautionary Principle forces one to estimate and to project catches and biomasses, even if they are not very precise. This will be discussed later.

CHAPTER 2 – MODELS AND RATES

2.1 MODELS

Science builds models or theories to explain phenomena. One observes *phenomena* and then looks for relations, causes and effects. Observations are made about the evolution of a magnitude (characteristics) with time (or with other characteristics) and possible causes (factors) are explored. Examples:

- Physics – *phenomenum of the movement of the bodies* (characteristics – distance related to time spent)
- Biology – *phenomenum of growth* (characteristic – length or weight, related to time).

2.1.1 STRUCTURE OF A MODEL

Basic assumptions

The assumptions to serve as a basis for a model should :

- simplify reality
- be simple and mathematically treatable (manageable)
- not be contradictory
- not be demonstrated
- be established with the characteristics

Usually basic assumptions are related to the evolution of the characteristics. So, they are established on the variation rate of those characteristics and they do not need to be proved.

Relations (properties)

- they are deduced from the basic assumptions by the laws of logic (mathematics). The properties are also designated by

"results" or "conclusions" of the model.

Verification

- the results of the model must be coherent (to agree) with reality.

This implies the application of statistical methods and sampling techniques to check the agreement of the results with the observations.

Improvement

- if agreement is approximate, it is necessary to see if the approximation is enough or not.
- if the results do not agree with reality, then the basic assumptions have to be changed
- the changes can aim to the application of the model to other cases.

Advantages

- it is easier to analyse the properties of the model than the reality.
- the models produce useful results.
- they allow analysis of different situations or scenarios by changing values of the factors.
- to point out the essential parts of the phenomenon and its causes.
- they can be improved in order to adjust better to the reality.

2.1.2 SOME TYPES OF MODELS USED IN STOCK ASSESSMENT

Production Models

The production models are also designated as General Production models, Global models, Synthetical models or Lotka-Volterra type models. These models consider the stock globally, in particular the total abundance (in weight or in number) and study its evolution, the relation with the fishing effort, etc.. They do not consider the structure of the stock by age or by size.

Structural Models

These models consider the structure of the stock by age and the evolution of the structure with time. They mainly recognize that the stock is composed of individuals of different cohorts, and, consequently, of different ages and sizes. So, they analyse and they project the stock and the catches for the coming years, by following the evolution of its different cohorts.

This manual will not follow the chronological construction of the models. It was thought to be more convenient to deal firstly with the structural models and afterwards with the production models.

2.2 RATES

The basic assumptions of a model, for the evolution of a characteristic, require the concept of variation rate of the characteristic related to time (or to other characteristics).

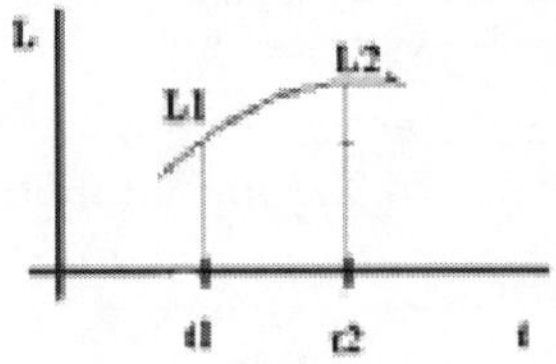

Figure 2.1 Evolution of the length (L) of an individual with time (or age) (t)

In order to generalize the study of the rates, the characteristic L in the example above will be substituted by y, and the associated variable will not be time, t, but the variable x. To study the stock assessment models and to make this study easier, it will be considered that the function y will only assume real and positive values.

2.2.1 ABSOLUTE MEAN RATE – amr (y)

Consider y a function of x and the interval i with the limits (x_i, x_{i+1})

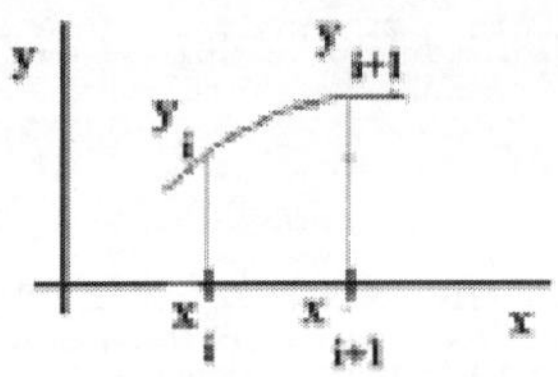

Figure 2.2 Function y= f(x) with variation in the interval i

Let :

$\Delta x_i = x_{i+1} - x_i$ be the size of the interval

y_i = the value of y when $x = x_I$

y_{i+1} = the value of y when $x = x_{i+1}$

The variation of y in the interval Δx_i will be $\Delta y_i = y_{i+1} - y_i$

The absolute mean rate, amr (y), of the variation of y within the interval Δx_i, will be :

$$\text{amr}(y) = \frac{\Delta y_i}{\Delta x_i}$$

Graphically :

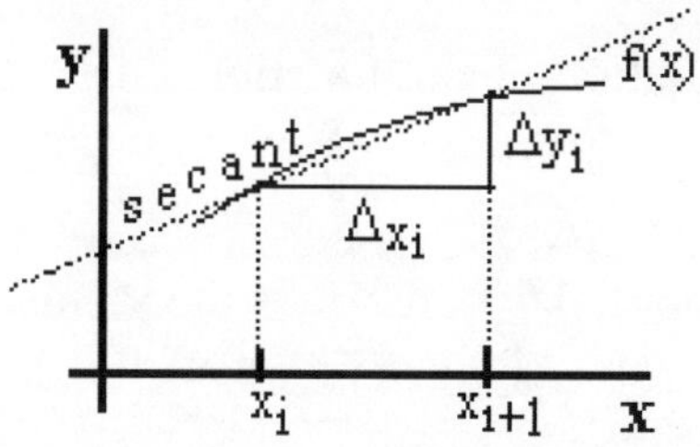

Figure 2.3 Absolute mean rate of the variation of y within the interval Δx_i

Slope of the secant = $\dfrac{\Delta y_i}{\Delta x_i}$ = amr (y) during Δx_i

Note: amr (y) is known in physics as the mean velocity of the variation of y with x, in the interval Δx_i.

2.2.2 ABSOLUTE INSTANTANEOUS RATE – air (y)

Let y be a function of x

The absolute instantaneous rate of y at the point x = x_i is the derivative of y in order to x at that point.

$$\text{air}(y)_{x=x_i} = \left(\frac{dy}{dx}\right)_{x=x_i}$$

Graphically :

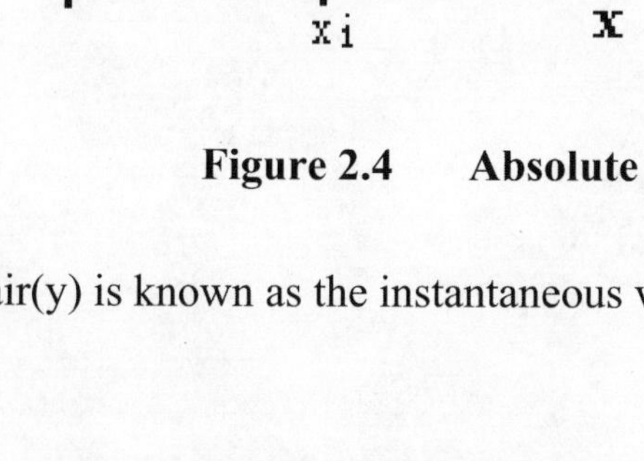

$\text{air}(y)_{x=x_i}$ equal to the slope of the tangent to the curve at the point $x=x_i$

Figure 2.4 Absolute instantaneous rate of y at point x_i

Note: air(y) is known as the instantaneous velocity of variation of y with x at the point x.

Properties

1. Given the value of air(y) the calculation of the function y is obtained, by integration, being $y = f(x) + Constant$, where $f(x)$ = Primitive of air (y) and Constant is the constant of integration.

   If the initial condition x^* , y^* is adopted, where y^* is the value of y corresponding to $x = x^*$, eliminating the Constant, then one can write $y = y^* + f(x) - f(x^*)$

2. The angle made by the tangent to the curve y with the xx's axis is designated by inclination.

   The trigonometric tangent of the inclination is the slope of the geometrical tangent.

   air (y) = derivative of y = slope = tg (inclination)

3. If, at point x :

   air (y) > 0 then y is increasing at that point

   air (y) < 0 then y is decreasing at that point

   air (y) = 0 then y is stationary at that point (maximum or minimum)

4. If air (y) is constant (= const) then y is a linear function. From property 1, it will be :

   $y = Constant + const.\ x$ or

   $y = y^* + const.(x - x^*)$ and vice versa

5. If $y(x) = u(x) + v(x)$ then air (y) = air(u) + air(v)

6. If factors A and B cause variations in **y**, then factors A and B considered *simultaneously* cause a variation of **y** with:

   $$air\ (y)_{total} = air\ (y)_{causeA} + air\ (y)_{causeB}$$

   $$air(air(y)) = \frac{d^2y}{d^2x} = \text{acceleration of y at the point x}$$

7. If the acceleration at the point x is increasing, then air (y) is positive and if that acceleration is decreasing, then air (y) will be negative.

2.2.3 RELATIVE MEAN RATE - rmr (y)

Consider y a function of x and the interval (x_i, x_{i+1})

Let :

$\Delta x_i = x_{i+1} - x_i$ = the size of the interval

y_i = value of y when $x = x_I$

y_{i+1} = value of y when $x = x_{i+1}$

x_i^* = a certain point in the interval (x_i, x_{i+1})

y_i^* = value of y when $x = x_i^*$

x_i^* can be x_i , x_{i+1} , $x_{central_i}$, etc.

The *mean rate of y relative to y_i^** will be :

$$rmr(y) = \frac{1}{y_i^*} . \frac{\Delta y_i}{\Delta x_i} \qquad \text{or}$$

$$rmr(y) = \frac{1}{y_i^*} .amr(y)$$

Comments

1. rmr (y) is associated with the mean rate of the variation of the percentage of y in relation to y*, that is:

$$\frac{\Delta(\frac{y_i}{y_i^*})}{\Delta x_i}$$

2. Let $\quad x_{central_i}$ be $x_{central_i} = x_i + \frac{\Delta x_i}{2} = \frac{1}{2}.(x_i + x_{i+1}) = \bar{x}_i$

3. It is convenient to designate by $y_{central_i}$ the value of y in the interval (x_i , x_{i+1}) when $x = x_{central_i}$.

   Notice that $y_{central_i}$ can be different from the mean, $\frac{(y_i + y_{i+1})}{2}$

4. It is frequent to calculate rmr(y) in relation to $y_{central_i}$ of the interval.

2.2.4 RELATIVE INSTANTANEOUS RATE - rir(y)

Let y be a function of x.
The relative instantaneous rate of y at the point $x = x_i$ is

$$rir(y) = \frac{1}{y_i} \cdot \left(\frac{dy}{dx}\right)_{x=x_i} \qquad \text{or} \qquad rir(y) = \frac{air(y)_{x=x_i}}{y_i}$$

Properties

1. Given rir(y), the calculation of the function y is obtained by integration, being

   y = f(x) + Constant, where f(x) = Primitive of rir(y) and C is the constant of integration.

   If one accepts the initial condition x* , y* , where y* is the value of y corresponding to x = x*, one will get, eliminating the Constant, $y = y^* + f(x) - f(x^*)$

2. If, at a point x :

rir (y) > 0	then y is increasing at that point
rir (y) < 0	then y is decreasing at that point
rir (y) = 0	then y is stationary at that point (maximum or minimum)

3. rir(y) = air (lny) as can be deduced from the derivation rules.

4. If rir (y) = constant = (const) then y is an exponential function of x, that is,

   $y = \text{Constant} \,.\, e^{\text{const}.x}$ or

   $y = y^* \,.\, e^{\text{const}.(x-x^*)}$ and vice-versa

5. If y(x) = u(x).v(x) then rir(y) = rir(u) + rir(v)

6. If the factors A and B cause variations in y, then simultaneously, factors A and B cause a variation in y, with:

   $$\text{rir}\ (y)_{\text{total}} = \text{rir}\ (y)_{\text{cause A}} + \text{rir}\ (y)_{\text{cause B}}$$

2.3 SIMPLE LINEAR MODEL

Let y = f(x)

Basic assumption of the model

air(y) = *Constant* = b in the interval (x_i, x_{i+1}) *with* $\Delta x_i = x_{i+1} - x_i$

Initial Condition

$x^* = x_i \Rightarrow y^* = y_i$

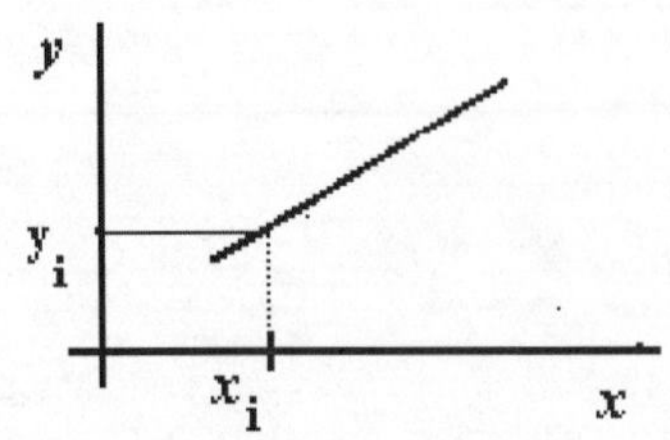

Figure 2.5 Graphical representation of a simple linear model

Properties

1. General expression $y = y_i + b \cdot (x - x_i)$; $y = a + bx$

2. Value y_{I+1} at the end of the interval, Δx_i $y_{i+1} = y_i + b \cdot \Delta x_i$

3. Variation, Δy_i , in the interval, Δx_i $\Delta y_i = y_{i+1} - y_i = b.\Delta x_i$

4. Central value, $y_{central_i}$ of the interval, Δx_i $y_{central_i} = y_i + b.(x_{centrali} - x_i) = y_i + b \cdot \frac{\Delta x_i}{2}$

5. Cumulative value, y_{cum_i} during the interval, Δx_i $y_{cum_i} = \int_{x_i}^{x_{i+1}} y.dx = \Delta x_i \cdot (a + b \cdot \bar{x}_i)$

   or from the Property 1

   $y_{cum_i} = \Delta x_i \cdot [y_i + b \cdot (\bar{x}_i - x_i)]$

6. Mean value, $\bar{y}_i$, in the interval, Δx_i $\bar{y}_i = \frac{y_{cum_i}}{\Delta x_i} = a + b \cdot \bar{x}_i$ where

   $\bar{y}_i = \frac{y_{cum_i}}{\Delta x_i} = y_i + b \cdot (\bar{x}_i - x_i)$

Other useful expressions

7. Cumulative value, y_{cum_i} during the interval, Δx_i $y_{cum_i} = \Delta x_i \cdot \bar{y}_i$

8. Mean value, $\bar{y}_i$, during the interval, Δx_i $\bar{y}_i = y_i + b.(\bar{x}_i - x_i)$ where $\bar{y}_i = a + b\bar{x}_i$

9. Mean value, $\bar{y}_i$, in the interval, Δx_i $\bar{y}_i = y_i + b\frac{\Delta x_i}{2}$

10. Mean value, $\bar{y}_i$, during the interval, Δx_i $\bar{y}_i = y_{central_i}$

11. Relation between amr(y) et air(y) $amr(y_i) = \frac{\Delta y_i}{\Delta x_i} = b = air(y)$

12. If $\Delta y_i < 0$ then $b < 0$ et vice-versa

13. In the linear model, the arithmetic mean of y_i and y_{i+1} is equal to the mean value, $\bar{y}_i$, and equal to the central value $y_{central_i}$

Important demonstrations

General expression Property 1

If tia(y) = b in the interval Δx_i then y is linear with x and considering the initial condition it will be: $y = y_i + b.(x-x_i)$

Central value Property 4

$$y_{central_i} = y_i + b.(x_{centrali} - x_{centrali}) = y_i + b\left((x_i + \frac{\Delta x_i}{2}) - x_i\right) = y_i + b.\frac{\Delta x_i}{2}$$

Cumulative value Property 5

from the definition of the cumulative value :

$$y_{cum_i} = \int_{x_i}^{x_{i+1}} (a + bx) \cdot dx$$

$$= a(x_{i+1} - x_i) + b\left(\frac{x_{i+1}^2}{2} - \frac{x_i^2}{2}\right)$$

it will be necessary to use the factorization of the difference of two squares, that is :

$$x^2_{i+1} - x^2 = (x_{i+1} - x_i).(x_{i+1} + x_i) = \Delta x_i \,.\, (x_{i+1} + x_i)$$

and then:

$$y_{cum\,i} = a\Delta x_i + b\Delta x_i \,.\, \bar{x}_i = \Delta x_i\,(a + b.\ \bar{x}_i)$$

$\bar{y}_i$ et $y_{central_i}$ Property 10

$$\bar{y}_i = y_i + b.(\bar{x}_i - x_i) = y_i + b\left(\frac{x_{i+1}}{2} - \frac{x_i}{2}\right) = y_i + b.\frac{\Delta x_i}{2} = y_{centrali}$$

2.4 EXPONENTIAL MODEL

Let y = f(x)

Basic assumption of the model

rir(y)=Constant=c in the interval (x_i, x_{i+1}), with $\Delta x_i = x_{i+1} - x_i$

Initial condition

$$x^* = x_i \Rightarrow y^* = y_i$$

Properties

rir(y) = air(lny) means that the exponential model of y against x is equivalent to the linear model of lny against x. So being, its properties can be deduced by backwards application of logarithm rules to the properties of the linear model of lny against x.

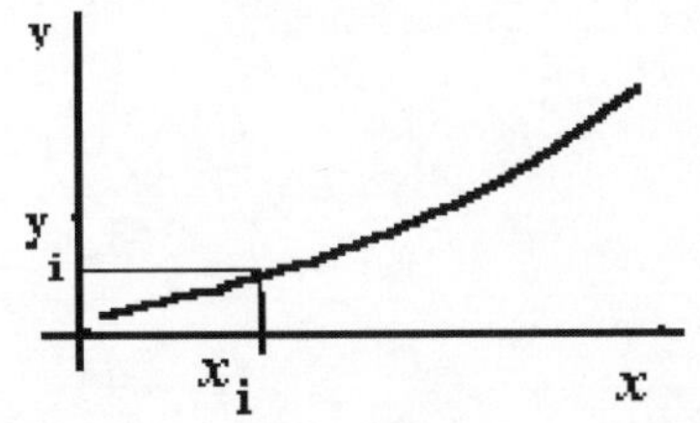

Figure 2.6 Graphical representation of the exponential model

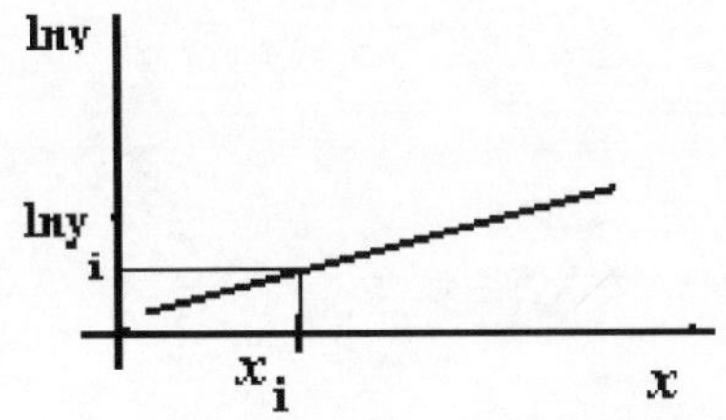

Figure 2.7 Graphical representation of the linear model of lny against x

		Exponential model of y (y against x) ←	Linear model lny (lny against x)
1.	General expression	$y = y_i \cdot e^{c \cdot (x - x_i)}$ ←	$\ln y = \ln y_i + c(x - x_i)$
2.	Value of y_{i+1} at the end of the interval, Δx_i	$y_{i+1} = y_i \cdot e^{c \cdot \Delta x_i}$ ←	$\ln y_{i+1} = \ln y_i + c\Delta x_I$
3.	Variation, Δy_i , during the interval, Δx_I	$\Delta y_i = y_{i+1} - y_i = y_i \cdot (e^{c \cdot \Delta x_i} - 1)$ calculated from 1	
4.	Central value, $y_{central_i}$, in the interval Δx_i	$y_{central_i} = y_i . e^{\frac{c . \Delta x_i}{2}}$ ←	$\ln y_{central_i} = \ln y_i + \frac{c\Delta x_i}{2}$
		$y_{central_i} = (y_i . y_{I+1})^{1/2}$ ←	$\ln y_{central_i} = (\ln y_i + \ln y_{i+1})/2$
		($y_{central_i}$ = geometric mean of the extremes y_i and y_{i+1})	

5. Cumulative value, y_{cum_i}, during the interval, Δx_i

$$y_{cum_i} = \int_{x_i}^{x_{i+1}} y.dx = \frac{\Delta y_i}{c}$$

6. Mean value, $\overline{y_i}$, during the interval, Δx_I

$$\overline{y}_i = \frac{y_{cum_i}}{\Delta x_i} = \frac{1}{c}.\frac{\Delta y_i}{\Delta x_i}$$

$$\overline{y}_i = y_i.\frac{e^{c\Delta x_i} - 1}{c\Delta x_i}$$ (replacing Δy_i using Propriety 3)

$$\overline{y}_i = \frac{y_{i+1} - y_i}{\ln y_{i+1} - \ln y_i}$$

$$\overline{y_i} \approx y_{central_i}$$

Other useful expressions

7. Expressions of variation, Δy_i

$$\Delta y_i = c.\ y_{cum_i}$$

$$\Delta y_i = c.\overline{y}_i \Delta x_i$$

8. Expression of amr (y)

$$amr(y) = \frac{\Delta y_i}{\Delta x_i} = c.\overline{y}_i$$

9. Expression of rmr (y) in relation to $\overline{y_i}$

$$rmr(y)_{\text{in relation to } \overline{y_i}} = c = tir(y)$$

10. Expression of rmr (y)

$$rmr(y) = amr\ (\ln y) = \frac{\Delta \ln y_i}{\Delta x_i} = c$$

11.

y decreases

If $\Delta y_i < 0$ alors
$$c < 0$$
$$amr(y) < 0$$
$$rmr(y) < 0$$
$$\begin{pmatrix} y_{cum_i} > 0 \\ \overline{y}_i > 0 \end{pmatrix}$$
and vice-versa

12. In the exponential model, the geometric mean of y_i and y_{i+1} is equal to the central value, $y_{central_i}$ (Prop. 4) and approximately equal to the mean value, $\overline{y}_i$ (Prop. 6), been the approximation better when Δx_i is smaller.

Demonstrations

Cumulative value
Property 5

$$y_{cum_i} = \int_{x_i}^{x_{i+1}} y \cdot dx = \int_{x_i}^{x_{i+1}} y_i \cdot e^{c\cdot(x-x_i)} \cdot dx = \cdot \left[\frac{y}{c}\right]_{x_i}^{x_{i+1}} = \frac{1}{c} \cdot \Delta y_i$$

Relation between $\overline{y}_i$
and $y_{central_i}$

Property 6 –
4th expression

From the approximation $\frac{e^h - 1}{h} \approx e^{h/2}$ with $h = c.\Delta x_i$

and from property 6-2nd expression, will be:

$$\overline{y_i} \approx y_i \cdot e^{c\Delta x_i/2}$$

Finally, by property 4-1st expression, one can conclude that:

$\overline{y}_i \approx y_{central_i}$

CHAPTER 3 – COHORT

3.1 COHORT – INTRODUCTION

A cohort or annual class or a generation, is a group of individuals born in the same spawning season. The following scheme illustrates the different phases of the life cycle of a cohort:

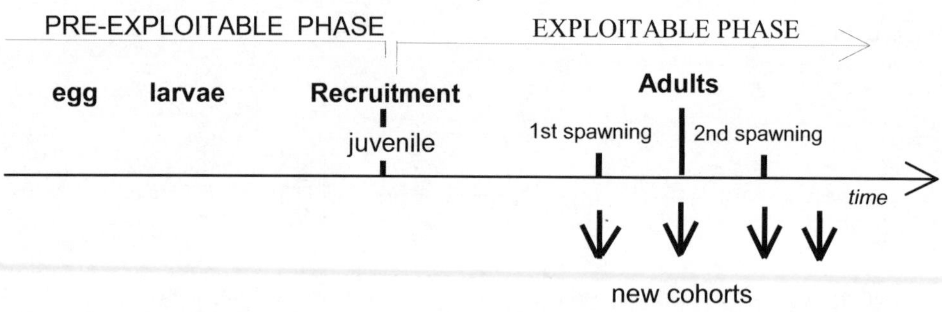

Figure 3.1 Cycle of life of a cohort

Let us start, for example, with the egg phase. The phases that follow will be larvae, juvenile and adult.

The number of individuals that arrive in the fishing area for the first time is called recruitment to the exploitable phase. These individuals grow, spawn (once or several times) and die.

After the first spawning the individuals of the cohort are called adults and in general, they will spawn again every year, generating new cohorts.

The phases of life of each cohort which precede the recruitment to the fishing area (egg, larvae, pre-recruits), are important phases of its life cycle but, during this time they are not usually subjected to exploitation. The variations in their abundances are mainly due to predation and environmental factors (winds, currents, temperature, salinity,…). *In these non exploitable phases mortality is usually very high, particularly at the end of the larvae phase* (Cushing, 1996). *This results in a small percentage of survivors until the recruitment. Notice that this mortality is not directly caused by fishing.*

The recruitment of a cohort during the exploitable phase, may occur during several months in the following schematic ways :

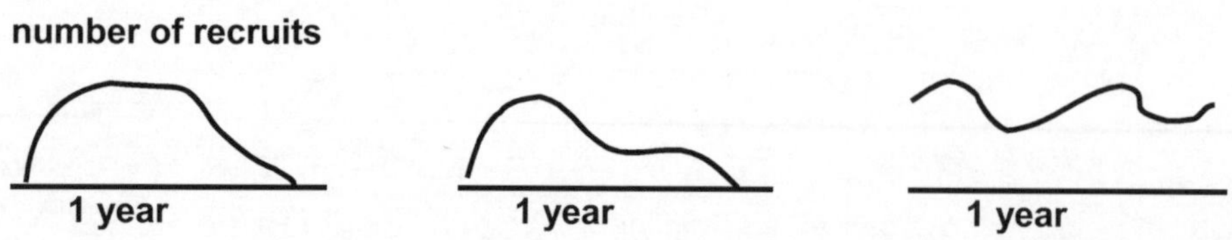

Figure 3.2 Types of annual recruitment to the exploitable phase

With some exceptions, the forms of recruitment can be simplified by considering that all the individuals are recruited at a certain instant, t_r called age of recruitment to the exploitable phase. It was established that recruitments will occur on 1 January (beginning of the year in many countries). These two considerations do not usually change the results of the analyses, but simplify them and agree with the periods of time to which commercial statistics are referred.

It should be mentioned that not all the individuals of the cohort spawn for the first time at the same age. The proportion of individuals which spawn increases with age, from 0 to 100 percent. After the age at which 100 percent of the individuals spawned for the 1st time, all the individuals will be adult. The histogram or curve that represents these proportions is called maturity ogive.

In certain cases, the maturity ogive can also be simplified supposing that the 1st spawning occurs at the age t_{mat} designated as age of 1st maturity. This simplification means that the individuals with an age inferior to t_{mat} are considered juveniles and those with the same age or older, are considered adults.

Figure 3.3 represents a maturity ogive with the shape of a histogram or curve :

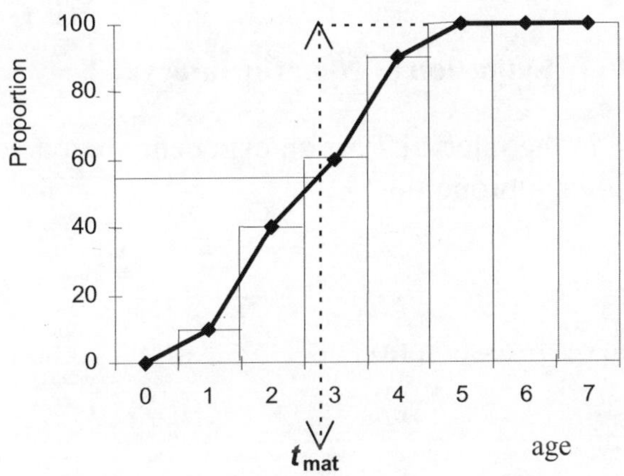

Figure 3.3 Maturity ogive

3.2 EVOLUTION OF THE NUMBER OF A COHORT, IN AN INTERVAL OF TIME

Consider the interval (t_i, t_{i+1}) with the size $T_i = t_{i+1} - t_i$ of the evolution of a cohort with time and N_t the number of survivors of the cohort at the instant t in the interval T_i (see Figure 3.4).

The available information suggests that the mean rates of percentual variation of N_t can be considered approximately constant, that is, rmr $(N_t) \approx$ constant.

Basic assumption

The relative instantaneous rate of variation of N_t, in the interval T_i is :

$$rir\,(N_t) = \text{constant negative} = -\,Z_i$$

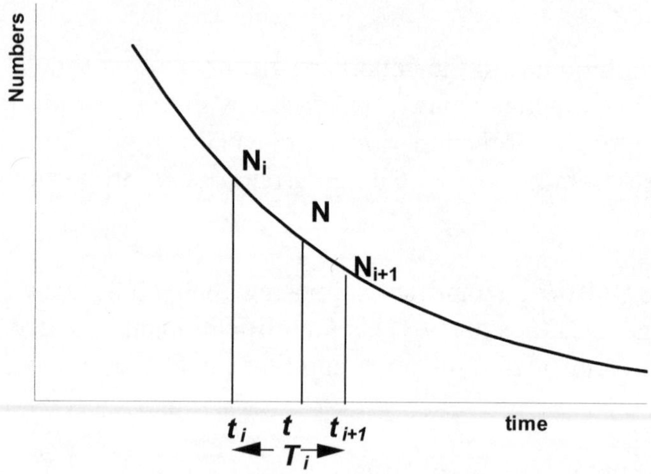

N_i = number of survivors at the beginning of the interval (t_i, t_i +1)
N_{i+1} = number of survivors at the end of the interval (t_i, t_i +1)

Figure 3.4 Evolution of N in the interval T_i

The model of the evolution of N_t, in the interval T_i, is an exponential model (because rir(N_t is constant). This model has the following properties:

Properties

1. General expression. From the basic assumption

$$rir(N_t) = -Z_i$$

with the initial condition that, for $t = t_i$ it will be $N_t = N_i$ then:

$$N_t = N_i \cdot e^{-Z\cdot(t-t_i)}$$

2. Number of survivors, N_{i+1}, at the end of interval T_i

$$N_{i+1} = N_i \cdot e^{-Z\cdot T_i}$$

3. Number of deaths, D_i, during the interval T_i

$$D_i = N_i - N_{i+1}$$

$$D_i = N_i(1 - e^{-Z_i T_i})$$

(notice that D_i is positive but the variation $\Delta N_i = N_{i+1} - N_i$ is negative)

4. Cumulative number of survivors, N_{cumi}, during the interval T_i

$$N_{cumi} = \frac{D_i}{Z_i}$$

$$N_{cumi} = N_i \cdot \frac{1 - e^{-Z_i \cdot T_i}}{Z_i}$$

5. Approximate central value, $N_{centrali}$, in the interval T_i

$$N_{centrali} \approx N_i \cdot e^{-Z_i \cdot T_i/2}$$

6. Mean number, $\overline{N_i}$, of survivors during the interval T_i

$$\overline{N_i} = \frac{N_{cumi}}{T_i}$$

$$\overline{N_i} = N_i \cdot \frac{1 - e^{-Z_i \cdot T_i}}{Z_i \cdot T_i}$$

$$\overline{N_i} = \frac{D_i}{Z_i \cdot T_i}$$

$$\overline{N_i} = \frac{N_i - N_{i+1}}{\ln N_i - \ln N_{i+1}} \qquad \text{(Ricker)}$$

$$\overline{N_i} \approx N_i \cdot e^{-Z_i \cdot T_i/2}$$

$$\overline{N_i} \approx N_{centrali} \quad \text{when} \quad \frac{Z_i \cdot T_i}{2} \quad \text{is small } (Z_i.T_i < 1)$$

Comments

1. The basic assumption is sometimes presented in terms of absolute instantaneous rates, that is:

   air $(N_t) = - Z_i .N_t$ { air (N_t) proportional to N_t} or

   air $(\ln N_t) = -Z_i$

   Z_i = mortality total coefficient, assumed constant at the interval T_i

   Notice that:

   $+Z_i$ = rir of total mortality of N_t

   $-Z_i$ = rir of variation of N_t

2. Unit of Z_i

From the definition, it can be deduced that Z_i is expressed in units of $[time]^{-1}$. By agreement, the unit $year^{-1}$ has been adopted, even when the interval of time is smaller or bigger than a year.

The following expressions show, in a simplified way, the calculation of the unit of Zi, with the rules and usual symbols [..] of dimension in the determination of physical units .

$$\frac{[1]}{[N_t]} \cdot \frac{[dN_t]}{[dt]} = [-Z_i]$$

$$\frac{1}{number} \cdot \frac{number}{time} = +[Z_i] \qquad \text{then} \qquad [Z_i] = time^{-1}$$

3. Annual survival rate, S_i

When $T_i = 1$ year, it will be :

$$N_{cumi} = \overline{N}_i = \frac{D_i}{Z_i} \qquad \text{and also}$$

$$N_{i+1} = N_i \cdot e^{-Z_i}$$

S_i = Annual survival rate in the year i
(or *percentage of the initial number of individuals that survived at the end of the year*).

$$S_i = \frac{N_{i+1}}{N_i}$$

$$S_i = e^{-Z_i}$$

1 - S_i = Annual mortality rate in the year i

The percentage of the initial number of individuals that die during the year is, by definition, the relative mean rate rmr (N_t) of mortality of N_t , over one year, in relation to the initial number, N_I

$$1 - S_i = \frac{Di}{Ni} = 1 - \frac{N_{i+1}}{N_i} = \frac{N_i - N_{i+1}}{N_i}$$

$$1 - S_i = 1 - e^{-Z_i}$$

4. Absolute mean rate $\qquad amr(N_t) = -Z_i.\overline{N}_i$

5. Relative mean rate $\qquad rmr(N_t) = -Z_i$ in relation to $\overline{N}_i$

6. Notice that S_i takes values between 0 and 1 , that is:

$$0 \leq S_i \leq 1 \qquad \text{but} \qquad Z_i \text{ can be} > 1$$

7. If the limits of the interval T_i were (t_i, ∞) then it would be:

$$T_i = \infty$$

$$N_{i+1} = 0$$

$$D_i = N_i$$

$$N_{cum_i} = \frac{N_i}{Z_i} \quad \text{and}$$

$$\overline{N}_i = 0$$

3.3 CATCH, IN NUMBER, OVER AN INTERVAL OF TIME

The causes of death of the individuals of the cohort due to fishing will be separated from all other causes of death. These other causes are grouped together as one cause designated as natural mortality. So, from the properties of the exponential model, the result will be

$$rir(N_t)_{\text{ total}} = rir(N_t)_{\text{ natural}} + rir(N_t)_{\text{ fishing}}$$

Supposing that, in the interval T_i, the instantaneous rates of mortality due to natural causes and due to fishing are constant and equal to M_i and to F_i, respectively, then

$$Z_i = F_i + M_i$$

Multiplying both factors of the previous equation (equality) by $.N_{cum_i}$, then:

$$Z_i \cdot N_{cum_i} = F_i \cdot N_{cum_i} + M_i \cdot N_{cum_i}$$

$Z_i.N_{cum_i}$ is the number, D_i , of deaths due to total mortality,

$$D_i = Z_i.N_{cum_i}$$

In the same way $F_i.N_{cum_i}$ will be the number of dead individuals due to fishing, that is, the Catch, C_i, in number, and then:

$$C_i = F_i.N_{cum_i}$$

Notice too that, $M_i.N_{cum_i}$ will be the number of dead individuals due to "natural" causes.

The exploitation rate, E_i , during the interval T_i was defined by Beverton and Holt (1956) as:

$$E_i = \frac{\text{number capture}}{\text{number dead}} = \frac{C_i}{D_i}$$

and then, $$E_i = \frac{F_i.N_{cum_i}}{Z_i.N_{cum_i}} = \frac{F_i}{Z_i}$$ or

$$E_i = \frac{F_i}{F_i + M_i}$$

The capture in number, C_i, in the interval T_i, can be expressed in the following different ways:

$$C_i = F_i.N_{cum_i}$$

$$C_i = F_i . \bar{N}_i . T_i$$

$$C_i = E_i.D_i$$

$$C_i = \frac{F_i}{Z_i}.D_i$$

$$C_i = \frac{F_i}{F_i + M_i} \cdot N_i\left[1 - e^{-(F_i + M_i)T_i}\right]$$

Comments

1. Ricker (1975) defines the exploitation rate, Ei*, as the percentage of the initial number that is captured in the interval Ti, that is: $E_i^* = C_i / N_i$.

    a) Ricker's definition may be more natural, but mathematically Beverton & Holt's definitions are more useful.

    b) It is easy to verify that $E_i^* = E_i.(1 - e^{-Z_i.T_i})$

2. The exploitation rate, E_i, does not have any unit, it is an abstract number.

3. The possible values of E_i are between 0 e 1, being 0 when there is no exploitation and 1 when the capture C_i is equal to the number of total deaths D_i, that is, when $M_i = 0$.

3.4 INDIVIDUAL GROWTH

In order to study the evolution of the biomass of a cohort, one can use the model of the evolution of a cohort, in number, and combine it with a model of the evolution of the mean weight of an individual of the cohort. In effect, the biomass B_t is equal to $N_t.W_t$ where W_t is the individual mean weight at the instant t.

To define a model for the individual growth weight W_t, there are then two possibilities :

ALTERNATIVE 1:

A) To define a model for the mean individual growth in length, L_t

B) To define the relation Weight-Length.

C) To combine A) with B) and obtain a mode for the mean individual growth in weight, W_t

ALTERNATIVE 2:

D) To define directly growth models for W_t and L_t.

ALTERNATIVE 1

A) Model for the individual growth by length

The models that are used in fisheries biology are valid for the exploitable phase of the resource. The most well known is the **von Bertalanffy Model** (1938) adapted by Beverton and Holt (1957). The existing observations suggest that there is an asymptotical length, that is, there is a limit to which the individual length tends.

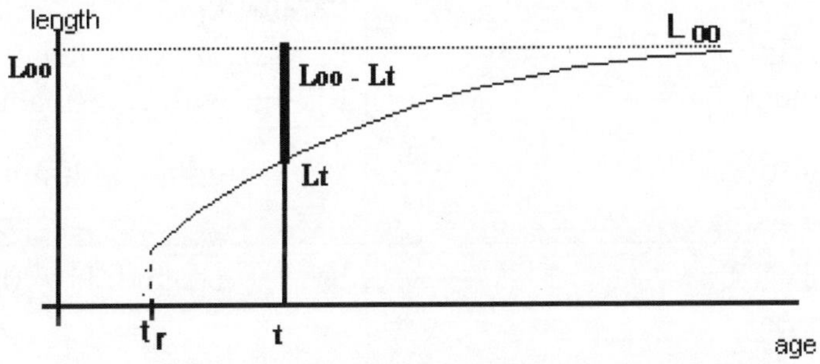

Figure 3.5 von Bertalanffy Model

t - age
L_t - individual mean length at the age t
L_∞ - asymptotical length
t_r – beginning of the exploitable phase

So, L_t presents an evolution where :

air(L_t) is not constant (because growth is not linear)

rir(L_t) is not constant (because growth is not exponential)

However, it can be observed that the variation of the quantity (L_∞ - L_t) (which we could call "what is left to grow"), presents a constant relative rate and can be described by an exponential model. So, we can adopt the:

Basic assumption

$$\text{rir}(L_\infty - L_t) = -K = \text{negative constant during all the exploited life}$$

where K is the growth coefficient (attention: the growth coefficient K is not the velocity of growth but the relative velocity of what "is left to grow" !!).

The properties of this model can be obtained directly from the general properties of the exponential model. The initial condition :

$$t = t_a \Rightarrow L_t = L_a$$

where t_a (and L_a), that corresponds to an instant within the exploitable phase, will be adopted.

The properties of the model of individual growth by length by Beverton & Holt (1957) deduced from the exponential model of $(L_\infty - L_t)$ are summarized as :

Properties of the exponential model for $(L_\infty - L_t)$	**von Bertalanffy Model for L_t**
1. **General expression**	
$(L_\infty - L_t) = (L_\infty - L_a)\cdot e^{-K\cdot(t-t_a)}$ Parameters L_∞ = asymptotic length K = growth coefficient $\begin{cases} t_a \\ L_a \end{cases}$ = initial condition	$L_t = L_\infty - (L_\infty - L_a)\cdot e^{-K\cdot(t-t_a)}$ Parameters L_∞ = asymptotic length K = growth coefficient $\begin{cases} t_a \\ L_a \end{cases}$ = initial condition
For $t_a = t_0$ and $L_a = 0$: $(L_\infty - L_t) = L_\infty \cdot e^{-K\cdot(t-t_0)}$	For $t_a = t_0$ and $L_a = 0$: $L_t = L_\infty - L_\infty \cdot e^{-K\cdot(t-t_0)}$ $L_t = L_\infty.\left[1 - e^{-K.(t-t_0)}\right]$
2. **Value at the end of the interval T_i**	
$(L_\infty - L_{i+1}) = (L_\infty - L_i)\cdot e^{-KT_i}$	$L_{i+1} = L_\infty \cdot (1 - e^{-KT_i}) + L_i \cdot e^{-K\cdot T_i}$ Ford-Walford expression (1933-1946)

3. Variation during the interval T_i	
$\Delta(L_\infty - L_i) = (L_\infty - L_{i+1}) - (L_\infty - L_i)$ $= (L_\infty - L_i).e^{-k.T_i} - (L_\infty - L_i)$ $= (L_\infty - L_i)\cdot\left[e^{-k.T_i} - 1\right]$	As $\Delta(L_\infty - L_i) = -\Delta L_i$ It will be:: $\Delta L_i = (L_\infty - L_i)\cdot(1 - e^{-K\cdot T_i})$
4. Cumulative value during the interval T_i	
$(L_\infty - L_i)_{cum_i} = \dfrac{\Delta(L_\infty - L_i)}{-K} = \dfrac{\Delta L_i}{K}$	$L_{cum_i} = L_\infty.T_i - \dfrac{\Delta L_i}{K}$ from: $(L_\infty - L_i)_{cum_i} = L_\infty.T_i - L_{cum_i} = \dfrac{\Delta L_i}{K}$
5. Mean value during the interval *T_i*	
$\overline{(L_\infty - L_i)} = \dfrac{(L_\infty - L_i)_{cum_i}}{T_i} = \dfrac{\Delta L_i}{K\cdot T_i}$	$\overline{L}_i = L_\infty - \dfrac{\Delta L_i}{K\cdot T_i}$ Gulland e Holt expression (1959) from: $\overline{(L_\infty - L_i)} = L_\infty - \overline{L}_i$
6. Central value of the interval *T_i*	
$(L_\infty - L)_{central_i} = (L_\infty - L_a)\cdot e^{-K\cdot\left(t_{central_i} - t_a\right)}$ and $(L_\infty - L)_{central_i} \approx \overline{(L_\infty - L_i)}$	$L_{central_i} = L_\infty - (L_\infty - L_a)\cdot e^{-K(t_{central_i} - t_a)}$ and $L_{central_i} \approx \overline{L_i}$

B) Relation Weight-Length

It is common to use the power function to relate the individual weight to the total (or any other) length. Then :

$$W_t = a\cdot L_t^{\,b}$$

where constant *a* is designated as condition factor and constant *b* as allometric constant. This relation can be justified accepting that the percentage of growth in weight is proportional to the percentage of growth in length, otherwise, the individuals would become disproportionate. Thus, the basic assumption is :

$$\text{rir}(W) = b\cdot \text{rir}(L)$$

where *b* is the constant of proportionality.

C) Combination of A) and B) and comments:	
1.	From the combination of $W = a \cdot L^b$ with $L_t = L_\infty \cdot (1-e^{-K(t-t_0)})$ we have $$W_t = W_\infty \cdot (1-e^{-K(t-t_0)})^b \text{ with } W_\infty = a \cdot L_\infty^b$$ This relation of growth in weight is designated as the Richards equation (1959). When b=3 the equation is the von Bertalanffy growth equation (1938).
2.	From, $W = a \cdot L^b$ we have, by definition, $W_{central_i} = a \cdot L^b_{central_i}$, where $W_{central_i}$ is the value corresponding to $L_{central_i}$
3.	Let $\overline{w}_i^*$ be the weight corresponding to $\overline{L}_i$, that is, $\overline{w}_i^* = a \cdot (\overline{L}_i)^b$ As, $\overline{L}_i \approx L_{central_i}$ then, $\overline{w}_i^* \approx a \cdot L^b_{central_i} = W_{central_i}$ In practice, $L_{central_i}$ and $W_{central_i}$ are preferred to the mean points
4.	The Richards and von Bertalanffy models are not the only models used for the evolution of W_t. Other models which have also been used in stock assessment are : Gompertz model (1825) and Ricker model (1969). (see Alternative 2 – property 3)
5.	Historically, the von Bertalanffy model was developed from the basic assumption $$\text{tia}(W_t) = Cte_1.W^{\frac{2}{3}} - Cte_2.W$$ Where Cte_1 and Cte_2 were designated by von Bertalanffy as the anabolism and the catabolism constants, respectively. Adopting the relation $W = a \cdot L^3$ the basic assumption will become $$\text{air}(L_t) = Cte_1 - Cte_2 . L$$ (where Cte_1 e Cte_2 are other constants). The solution of this diferential equation gives the von Bertalanffy equation.

ALTERNATIVE 2

D) Model directly for W_t and L_t

In alternative 1, a model was developed for growth in length, and then a model was built for growth in weight, using the relation between weight and length.

The basic assumption adopted by alternative 1 used the characteristic $(L_\infty - L_t)$ instead of weight in order to be able to have a characteristic with a rate rir constant, that is, an exponential model. The relation $W=a.L^b$ was adopted to obtain the model of growth in weight. Notice that it can be said that L was considered as a function of W, that is, $L=(W/a)^{1/b}$. It will then be possible to adopt, instead of that function of W, another function of the weight $H(W_t)$, in order to be able to formulate directly the basic assumption:

$$\text{air}[H(W_\infty)-H(W_t)]=-K=\text{constant}$$

with the initial condition

$$t=t_a \Rightarrow W_t=W_a$$

Properties

The properties of this model (once it is an exponential model) can be obtained directly from the general properties of the exponential model. It is particularly interesting to derive the general expression W_t resulting from different choices of function H.

1. **General expression** $[H(W_\infty)-H(W_t)]=[H(W_\infty)-H(W_a)].e^{-K(t-ta)}$

   or $H(W_t)=H(W_\infty)-[H(W_\infty)-H(W_a)]\cdot e^{-K\cdot(t-t_a)}$

2a. **Richards equation in weight** Adopting the following function $H(W_t)=W_t^{1/b}$

The result will be the general expression:

$$W_t^{\frac{1}{b}}=W_\infty^{\frac{1}{b}}-(W_\infty^{\frac{1}{b}}-W_a^{\frac{1}{b}})\cdot e^{-K(t-ta)}$$

that is, the Richards equation; and when b=3, this is the equation of von Bertalanffy, so:

2b. **von Bertalanffy equation in weight**, will be :

$$W_t^{\frac{1}{3}}=W_\infty^{\frac{1}{3}}-(W_\infty^{\frac{1}{3}}-W_a^{\frac{1}{3}})\cdot e^{-K(t-ta)}$$

3. **Gompertz equation in weight** Adopting the function $H(W_t)=\ln W_t$

The result will be the general expression:

$$\ln W_t=\ln W_\infty-(\ln W_\infty-\ln W_a)\cdot e^{-K(t-ta)})$$

4. The respective **equations in length** can be obtained by adopting other functions of $H(W_t)$:

4a. **von Bertalanffy equation in length** Adopting $H(W_t)=L_t$ it will be:

$$L_t=L_\infty-(L_\infty-L_a).e^{-K.(t-ta)}$$

4b. **Gompertz equation de in length** Adopting $H(W_t)=\ln L_t$ it will be:

$$\ln L_t=\ln L_\infty-(\ln L_\infty-\ln L_a)\cdot e^{-K.(t-ta)}$$

5. **Simplified equations**
The individual growth equations, both in length and in weight, are simplified when one selects $H(W_a) = 0$ for $t_a = t^*$

So, the simplified general expression will be reduced to : $H(W_t) = H(W_\infty).(1 - e^{-K\cdot(t-t^*)})$

5a. **Simplified Richards equation, in weight,** will be : $W_t = W_\infty \cdot \left[1 - e^{-K\cdot(t-t_0)}\right]^b$

where t^* was represented by t_0 because $H(W_a) = 0$ in Richards model means that W_a will also be zero.

5b. **Simplified Gompertz equation, in weight,** will be: $\ln W_t = \ln W_\infty \cdot \left[1 - e^{-K\cdot(t-t^*)}\right]$
In this case $H(W_a) = 0$ corresponds to $W_a = 1$

5c. **Simplified Richards equation in length,** will be: $L_t = L_\infty \cdot \left[1 - e^{-K\cdot(t-t_0)}\right]$
(with $L_a = 0$ for $t_a = t_0$)

5d. **Simplified Gompertz equation, in length,** will be: $\ln L_t = \ln L_\infty \cdot \left[1 - e^{-K\cdot(t-t^*)}\right]$
(with $L_a = 1$ for $t_a = t^*$)

Comments

1. Gompertz equation , in weight, is similar to Gompertz equation, in length, but, in their simplified forms, t* represents different ages, because they will correspond, respectively, to $W_a = 1$ and to $L_a = 1$.

2. Gompertz equation, in length, is similar to von Bertalanffy if L_t is substituted by $\ln L_t$. In practice, this fact allows the utilization of the same particular methods to estimate the parameters in both equations, using L in the von Bertalanffy expression and lnL in the Gompertz expression. (See Section 7.4)

3. It is important to notice, once again that, in practice, $L_{centrali}$ and $W_{centrali}$ are used instead of the mean values, $\overline{L_i}$ and $\overline{W_i}$.

4. Gompertz growth curve in length, has an inflection point, (t_{infl}, L_{infl}), with:

$$t_{infl} = t_a + (1/K).\ln(\ln(L_\infty/L_a)) \quad L_{infl} = L_\infty / e$$

5. Gompertz growth curve in weight has an inflection point, (t_{infl}, W_{infl}) with:

$$t_{infl} = t_a + (1/K).\ln(\ln(W_\infty/W_a)) \quad W_{infl} = W_\infty / e$$

6. Richards growth curve in length does not have an inflection point but the growth curve in weight has an inflection point, (t_{infl}, W_{infl}).

$$t_{infl} = t_a - \frac{1}{K}.\ln\left[\frac{1/b}{1-\left(\frac{W_a}{W_\infty}\right)^{1/b}}\right] \qquad W_{infl} = (1-\frac{1}{b})^b . W_\infty$$

In the particular case of the von Bertalanffy equation it will be :

$$W_{infl} = (8/27).W_\infty \qquad \text{and} \qquad t_{infl} = t_0+(1/k).\ln 3$$

7. Some authors refer Gompertz equation in other ways, for example, using the inflection point t_{infl} and the asymptotic weight W_∞, instead of the parameters t_a and W_a.

   It will then be $w_t = w_\infty.\exp\left(-e^{-k(t-t_{inf})}\right)$ or $L_t = L_\infty.\exp\left(-e^{-k(t-t_{inf})}\right)$
   Sometimes the length expression is presented in its general form: $L_t = a.\exp\left(b.e^{c.t}\right)$

   The parameters of the length model will then be: $L_\infty = a$; $k = -c$ et $t_{infl} = (1/c).\ln(-b)$

8. The growth in length presents an inflection in fish farming, where the study of growth covers very young ages and it is common to use the Gompertz equation. In fisheries, the tradition is to use the von Bertalanffy equation.

9. A model that can sometimes be useful, is the Ricker model (1975). This model is valid for a certain interval of time T_i and not necessarily for all the exploitable life of the fishery resource. In fact, the model is based on the basic assumption that the individual growth is exponential in the interval T_i.

It will be, for example, $L_t = L_i.e^{K_i.(t-t_i)}$ where K_i can be different from one interval to the next.

3.5 BIOMASS AND YIELD, DURING THE INTERVAL T_i

1. Biomass

Theoretically, it could be said that the biomass at the instant t of the interval T_i is given by:

$$B_t = N_t . W_t$$

Thus, the cumulative biomass during the interval T_i would be:

$$B_{cum_i} = \int_{t_i}^{t_{i+1}} B_t.dt$$

and the mean biomass in the interval T_i would be:

$$\bar{B}_i = \frac{B_{cum_i}}{T_i}$$

In the same way, the mean weight of the cohort, $\overline{W}_i$, in the interval T_i would be:

$$\overline{W}_i = \frac{B_{cum_i}}{N_{cum_i}}$$

The biomass can be obtained by dividing both terms of the fraction by T_i, as

$$\overline{B}_i = \overline{N}_i.\overline{W}_i$$

2. Yield

The yield, Y_i, during the interval T_i will be expressed as the product of the catch in numbers, times the individual mean weight :

$$Y_i = C_i \cdot \overline{W}_i$$

Comments

In practice $\overline{W}_i$ is considered approximately equal to $W_{central_i}$ at the interval T_i .

Other expressions of Y_i will also be :

$$\begin{aligned} Y_i &= F_i.N_{cum_i}.\overline{W}_i \\ &= F_i.B_{cum_i} \\ &= F_i.\overline{N}_i.\overline{W}_i.T_i \\ &= F_i.\overline{B}_i.T_i \end{aligned}$$

3.6 COHORT DURING THE EXPLOITABLE LIFE

Consider the evolution of a cohort during the exploitable life, beginning at age t_r, and intervals of time, T_i, covering all the exploitable phase (frequently the intervals are of 1 year...).

Figure 3.6 illustrates the evolution of the number of survivors of the cohort, N, and the evolution of the catches in number, C, which are obtained during the successive intervals of time T_i.

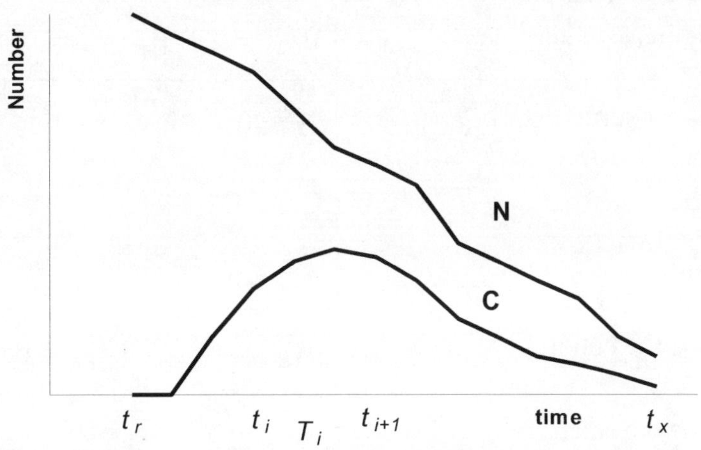

Figure 3.6 Evolution of the number of survivors of the cohort, N, and the catches in number, C

Figure 3.7 illustrates the evolution of the biomass of the cohort, B, and the evolution of the catches in weight, Y, which can be obtained during the successive intervals of time T_i.

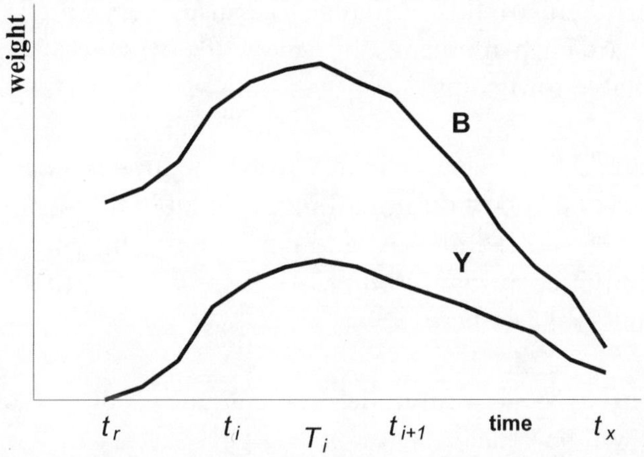

Figure 3.7 Evolution of the biomass of the cohort, B, and the catches in weight, Y

Values of the most important characteristics of the cohort, during all the exploitable phase

Duration of the life of the cohort	$\lambda = \Sigma T_i$
Total number of deaths	$D = \Sigma D_i$ (= R (recruitment) when all the individuals die)
Cumulative number of survivors	$N_{cum} = \Sigma N_{cum\,i}$
Mean number of survivors	$\overline{N} = N_{cum} / \lambda$

Cumulative biomass of survivors	$B_{cum} = \Sigma B_{cum\ i}$
Mean biomass of survivors	$\overline{B} = B_{cum} / \lambda$
Catch in number	$C = \Sigma C_I$
Catch in weight	$Y = \Sigma Y_I$
Mean weight of the cohort	$\overline{W}_{coorte} = B_{cum} / N_{cum} = \overline{B} / \overline{N}$
Mean weight of the catch	$\overline{W}_{capt} = Y / C$
Critical age	$t_{crítical}$ = age where B is maximum, when the cohort is not exploited.

Comments

1. At first sight it may seem that the values of the characteristics of a cohort during all the exploitable phase are of little interest, because, very rarely is fishing applied to an *isolated cohort.* At each moment, the survivors of several cohorts are *simultaneously* present and available for fishing.

2. Despite this fact and for reasons which will be mentioned later, it is important to analyse the characteristics of a cohort during all its exploitable life. Knowledge of the evolution of a cohort, in number and in biomass, and particularly the critical age, is important for the success of the activities in *fish farming.* As $B_t = N_t \,.\, W_t$, the critical age, $t_{crítical}$, will be the age t in the interval T_i where

   rir (W_t) = - rir(N_t)=M because, the critical age being the maximum biomass, the derivative of B will be equal to zero.

3. Notice that N_{cum} can be expressed in function of the recruitment as

   $$N_{cum} = R.\{...\}$$

   where {...} represents a function of biological parameters and annual fishing mortality coefficients F_i during the life of the cohort. B_{cum} can also be expressed as:

   $$B_{cum} = R\,.\{...\}$$

   where the function{...} also includes growth parameters.

3.7 SIMPLIFICATION OF BEVERTON AND HOLT

Beverton and Holt (1957) deduced algebraic expressions for the characteristics of a cohort during the exploited life, adopting the simple assumptions:

1. The exploited phase of the cohort is initiated at age t_c and is extended to the infinite.
2. The natural mortality coefficient, M, is constant during all the *exploitable phase*.

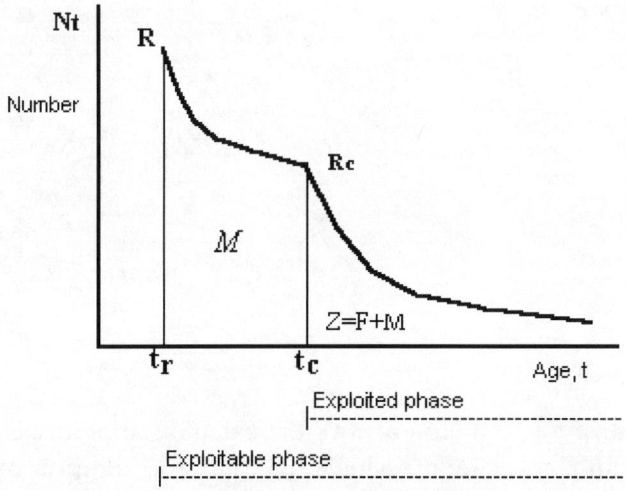

Figure 3.8 Life of a cohort during the exploitable phase

3. The fishing mortality coefficient, F, is constant during all the exploited phase.
4. Growth follows the von Bertalanffy equation with for $L_a = 0$ for $t_a = t_0$

Basic equations referring to the exploited phase

1. **c (ratio between length at age t_c and the asymptotical length)**

$$c = \frac{L_c}{L_\infty} = 1 - e^{-k\cdot(t_c - t_0)}$$

2. **Recruitment**

Y

$$R_c = R \cdot e^{-M(t_c - t_r)}$$

3. **Cumulative number**

$$N_{cum} = \frac{R_c}{Z}$$

4. **Catch in number**

$$C = E \cdot R_c$$

5. **Cumulative biomass**

$$B_{cum} = R_c \cdot W_\infty \cdot \left[\frac{1}{M+F} - 3\frac{(1-c)}{M+F+K} + 3\frac{(1-c)^2}{M+F+2K} - \frac{(1-c)^3}{M+F+3K}\right]$$

It can just be written $B_{cum} = R_c . W_\infty .[...]$

6. **Mean weight in the catch** $$\overline{W} = \frac{B_{cum}}{N_{cum}} = Z . W_\infty .[...]$$

7. **Catch in weight** $$Y = C \cdot \overline{W} = F . B_{cum} = F . R_c . W_\infty .[...]$$

8. **Mean age in the catch** $$\bar{t} = t_c + \frac{1}{Z}$$

9. **Mean length in the catch** $$\overline{L} = L_\infty - (L_\infty - L_c) . \frac{Z}{Z+K}$$

10. **Critical age** $$Y$$ $$t_{critique} = t_o - \frac{1}{K} . \ln\left(\frac{M}{M+3K}\right)$$

Comments

1. The simplification of Beverton and Holt allows the calculation of any characteristic of the cohort during its life with algebric expressions, avoiding the addition of the values of the characteristic in the successive intervals T_i. This was useful for calculations in the 60-70´s when computers were not available. It is also useful when the only available data is natural mortality, M, and growth parameters.

2. At present, the simplified expressions are also useful to study the effects on the biomass, on yield and on the mean weight of the catch due to changes in the fishing mortality coefficient, F, and/or on the age of first capture, t_c. These analyses are usually illustrated with figures. For example, Figure 3.9 exemplifies the analyses of the biomasses and of the catches in weight, obtained from a cohort during the exploited life, subjected to different fishing mortality coefficients, assuming a fixed age t_c .

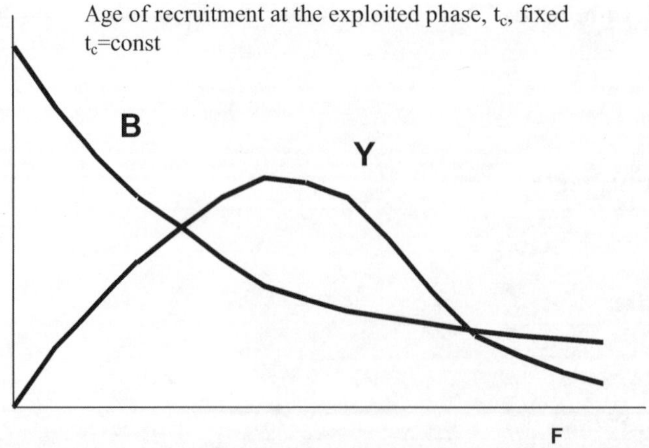

Figure 3.9 Evolution of the biomass and of the catch in weight from a cohort subject to different fishing mortality coefficients and fixed t_c (notice that the Figure illustrates only the analysis and does not take into consideration the scales of the axis).

3. Notice that the forms of the previous curves, Y and B_{cum} against F, do not depend on the value of the recruitment and so, they are usually designated as curves of biomass and yield per recruit, B/R and Y/R, respectively. The calculations are usually made with R=1000.

4. The mean weight, the mean age and the mean length of the catch do not depend on the value of the recruitment. The curves of the characteristics of a cohort during its life against the fishing level, F, or against the age of first catch, t_c, deserve a careful study, for reasons which will be stressed in the chapter concerning the long-term projections of the stock .

5. B_{cum} was calculated as:

$$B_{cum} = \int_{tc}^{\infty} N_t \cdot W_t .dt$$

where

$$N_t = R_c \cdot e^{-(M+F).(t-tc)} \quad \text{and} \quad W_t = W_\infty \cdot \left[1 - e^{-K.(t-to)}\right]^3$$

The calculations can also be made using other values of the constant b, from the relation W-L, different from the isomorphic coefficient, b=3, using the incomplete mathematical function Beta (Jones,1957).

6. The means $\bar{L}$ and $\bar{t}$ can be calculated from the cumulative expression

$$L_{cum} = \int_{tc}^{\infty} N_t \cdot L_t .dt \quad \text{and} \quad t_{cum} = \int_{tc}^{\infty} N_t .t.dt \quad \text{as}$$

$$\bar{L} = L_{cum} / N_{cum} \quad \text{and} \quad \bar{t} = t_{cum} / N_{cum}$$

These means are designated as *weighted means*, being the *weighting factors*, the number of survivals, N_t , at each age t.

CHAPTER 4 – STOCK

4.1 STOCK OVER A ONE YEAR PERIOD

4.1.1 EVOLUTION OF THE AGE STRUCTURE OF THE STOCK

Let us consider the evolution of a stock, with several cohorts, over the period of one year. Figure 4.1 shows the structure of the stock at the beginning of the year, the mean characteristics during the year and the age structure surviving at the end of the year.

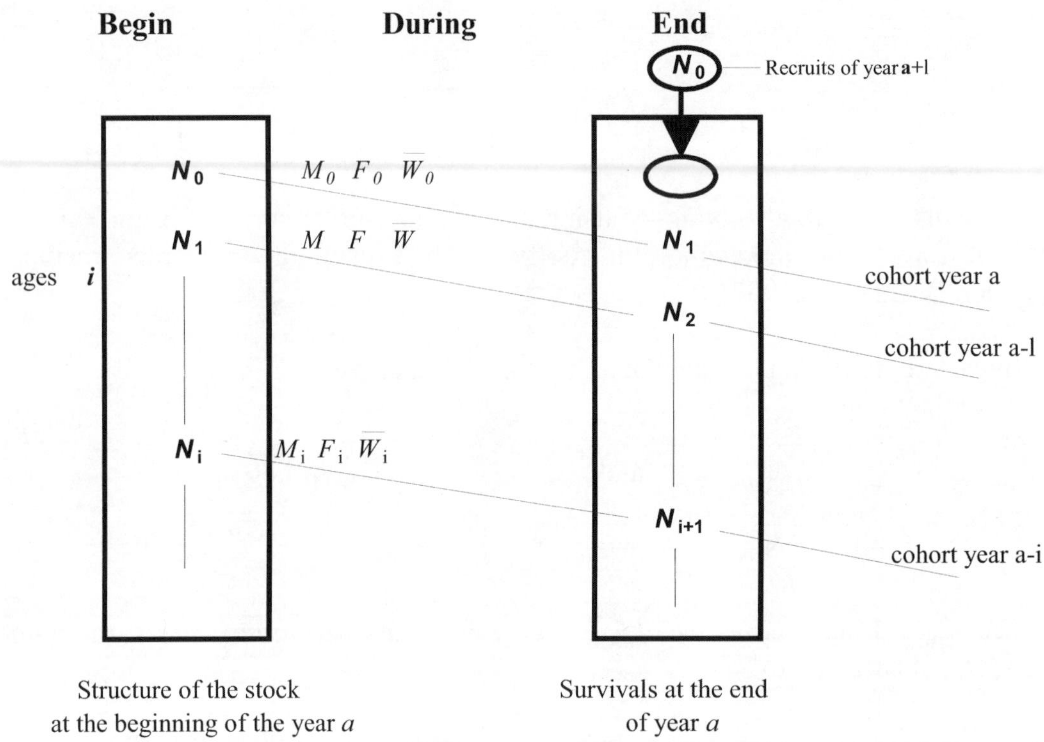

Figure 4.1 Structure of the stock at the beginning and end of the year

Let $i = 0,1,2,3, \ldots$ be the ages

4.1.2 CHARACTERISTICS OF THE STOCK AT THE BEGINNING OF THE YEAR

Let :

N_i – number of individuals at age i, at the beginning of the year.

w_i - individual weight at age i, at the beginning of the year.

B_i - total biomass of the individuals at age i, at the beginning of the year.

N - total number of survivors of the stock at the beginning of the year.

B - total biomass of the stock at the beginning of the same year.

Then :

$$N = \sum_i N_i$$

$$B = \sum_i B_i = \sum_i (N_i . W_i)$$

4.1.3 CHARACTERISTICS OF THE STOCK DURING THE YEAR

Let :

M_i, F_i e Z_i	total mortality (natural and by fishing) coefficients, at age **i** during the year
$\overline{W}_i$	individual mean weight at age i during the year
$\overline{N}$	mean number of individuals during the year
$\overline{B}$	mean biomass during the year
C	catch, in number, during the year
Y	catch, in weight, during the year
$\overline{W}_{catch}$	mean weight of the individuals caught during the year
$\overline{W}_{stock}$	mean weight of the individuals of the stock during the year

Then:

$$N_{cum} = \sum_i N_{cum\,i}$$

$$\overline{N} = \frac{N_{cum}}{T} \qquad \text{(where T=1 year)}$$

$$C = \sum_i C_i$$

$$B_{cum} = \sum_i B_{cum\,i}$$

$$\overline{B} = \frac{B_{cum}}{T} \qquad \text{(where T=1 year)}$$

$$Y = \sum_i Y_i$$

$$\overline{W}_{catch} = \frac{Y}{C}$$

$$\overline{W}_{stock} = \frac{\overline{B}}{\overline{N}}$$

with N_{cumi}, C_i, B_{cumi} and Y_i calculated for all the ages i according to the expressions given previously in Chapter 3 - Cohort.

4.1.4 CHARACTERISTICS OF THE STOCK AT THE END OF THE YEAR

The number of individuals at beginning of age i+1 will be:

$$N_{i+1} = N_i \cdot e^{-Z_i}$$

Let :

R - recruitment of the cohort in the following year

Then the number and the biomass of the stock at the beginning of the following year will be :

$$N = R + \sum_i N_{i+1} \qquad \text{and} \qquad B = R.W_1 + \sum_i B_{i+1}$$

where the product $R.W_1$ is the biomass of the recruitment of the following year.

Comments

1. The end of the year coincides with the beginning of the following year. So, the number of survivors of age *i* at the end of the year will be N_{i+1}, with age i+1.
2. The sum of the total number of survivors of the stock at the end of the year is not equal to the number of individuals of the stock at the beginning of the following year, because one has to count the recruits entering that year.
3. The total number of deaths, D, during the year, would be $D = \sum_i D_i$
4. As the interval of time is 1 year, the cumulative values will be equal to the mean values, that is :

$$N_{cum_i} = \overline{N}_i$$

$$B_{cum_i} = \overline{B}_i$$

$$\overline{N} = \sum_i \overline{N}_i$$

$$\overline{B} = \sum_i \overline{B}_i$$

5. The utilization of the same symbols N, B, D, etc., for the stock and for the cohort should not create any confusions.

4.2 FISHING PATTERN OVER A ONE YEAR PERIOD

4.2.1 FISHING LEVEL AND EXPLOITATION PATTERN

The direct action of fishing a stock can be represented by the coefficients of mortality by fishing F_i. These coefficients are associated with the quantity of effort, with the disponibility of the individuals of different sizes or ages, i, and with the fishing gears used by the vessels during the year.

It is usual to separate the coefficients of mortality by fishing into two components:
One is designated as the level of intensity of mortality by fishing, $\bar{F}$, during the year, called fishing level, F. The level is associated with the quantity of fishing effort, (the number of vessels fishing), the number of days, hauls, fishing hours during the year, and with the efficiency or the fishing power of the vessels or gears. Another component, designated as exploitation pattern, s_i, is associated with the selective properties of the fishing gears relative to the sizes or ages of the individuals available to be captured, during that year.

The combined set of the fishing level (a unique value for all ages) and the exploitation pattern (different values according to the size or the age), is designated as fishing pattern or fishing regime. The designation of fishing pattern may cause some confusion with the exploitation pattern, and the designation of fishing regime may be confused with what the economists and managers call fishing regime.

The fishing pattern, F_i, of one age i during one year, is equal to the product of the fishing level of that year, $\bar{F}$, times the exploitation pattern of each age, s_i. That is:

$$F_i = \bar{F}.s_i$$

To analyse the effects of the coefficient of mortality by fishing, $\bar{F}$, on the characteristics of the resource and on the catches, the exploitation pattern is generally taken as constant from one year to the next. Sometimes one analyses the effects of the changes of the exploitation pattern maintaining a fixed fishing level, but one can also analyse the combined effects of the two components.

The fishing level, $\bar{F}$, is often represented by F.

4.3 SHORT–TERM PROJECTIONS OF THE STOCK

Knowing the structure of the stock at the beginning of one year, it is possible to estimate the characteristics of the stock during that year and project the structure of the stock for the beginning of the next year (with an exception to the recruitment of that year), for different values of the fishing level, $\bar{F}$, (and for the exploitation pattern, s_i).

It is, of course, necessary to know the biological parameters of growth, maturity and the natural mortality coefficients, in each age during the year.

Adopting one value for the recruitment of the following year, the projection could be repeated for one more year and so on. The inconvenience in projecting the stock for several years will be that those projections will depend more and more on the adopted annual recruitment values. That is why, in practice, the stock and the catches are projected for one, or at the most, two years. In Section 4.5, the estimation of recruitments will be discussed.

Notice that to project a stock for the following year, it is necessary to have data from the previous year. So, the stock is firstly projected for the current year and then the catch and the biomass are projected for the following year.

Let us suppose, for example, that in 1997 one wants to project the characteristics of the stock for 1998. As the available data will be, in the best hypothesis, referring to 1996, one will have to project the stock of 1996 for the beginning of 1997 and together with the recruitment of 1997, project the stock until the end of 1997 and only after that, can one project to 1998, estimating previously the recruitment of 1998.

4.4 LONG–TERM PROJECTIONS OF THE STOCK

Let us consider the vector N with the components N_1, N_2, ..., N_i, ... representing the structure of the stock at the beginning of a year. Notice that N_1, N_2, ..., N_i, ... are the survivors of the different cohorts of the stock, at the beginning of the year.

Let M_i and F_i be the natural and fishing mortality coefficients of age i, assumed to be constant in the future years.

Figure 4.2 illustrates, with a theoretical stock, the projections of the numbers of survivors of the different cohorts at the beginning of 1980, for the years to come, from 1981-1986 (the values are expressed in million of individuals).

Notice that the recruitments are missing during the years 1981 to 1986, as they have not yet occurred. So, it is clear that the respective survivors are also missing during those years.

Let us suppose now that the recruitments of the same period of years were equal to those of 1980, that is, 440 million of individuals. Figure 4.3 shows the projections in future years. It can be seen that the values of the age structure of the stock in 1986 are equal to the annual evolution of the cohort in 1980.

Age	1980	1981	1982	1983	1984	1985	1986
0	440						
1	995	326					
2	367	680	223				
3	68	229	425	139			
4	245	41	139	258	85		
5	345	149	25	84	156	51	
6	76	209	90	15	51	95	31

Figure 4.2 Structure of the stock at the beginning of 1980 and projections of the number of survivors (million of individuals) of the different cohorts at the beginning of the years 1981-1986

Age	1980	1981	1982	1983	1984	1985	1986	1987
0	440	440	440	440	440	440	440	440
1	995	326	326	326	326	326	326	326
2	367	680	223	223	223	223	223	223
3	68	229	425	139	139	139	139	139
4	245	41	139	258	85	85	85	85
5	345	149	25	84	156	51	51	51
6	76	209	90	15	51	95	31	31

Figure 4.3 Long–term projections of the cohort of 1980 and structure of the stock assuming a constant annual recruitment of 440 million of individuals

One practical conclusion seems to be that to obtain the structure of the stock in 1986 it would be enough to follow the evolution of the cohort of 1980 and then, it would not be necessary to have the complete structure of the stock at the beginning of the year. It would be enough to adopt one value for the recruitment (R) of a cohort (and, of course, assume that the biological parameters would be stable in the following years). An advice would be to make the calculations adopting a *Recruitment of 1000 individuals* (with computer software a recruitment equal to 1 is adopted).

Except for the mean values – age, length, weight in the catch, etc, – which are independent of the adopted value of R, the characteristics of the stock in the long–term, under the previous conditions, are proportional to the recruitment. So, these projections are also designated as *projections by recruit* , by *LT projections* or even as *equilibrium projections*.

The long–term projections do not only concern the survivors at the beginning of the year, but also the values of the mean abundances, $\overline{N}_i$ during the year, and the catches in number, C_i .

One can also project the total biomasses, B or $\overline{B}$, the spawning biomasses, SB or $\overline{SB}$, as well as the catches in weight, Y, knowing the initial individual weights, W_i, the mean weights, $\overline{W}$, and other parameters like those of maturity.

It is important to verify that the cumulative values, $Ncum_i$ e $Bcum_i$, of a cohort during 1 year are equal to the mean values, $\overline{N}_i$ e $\overline{B}_i$, during that year. The long–term projections can be calculated as:

$$\overline{N}_{stock} = \sum Ncum_i = \sum \overline{N}_i$$

$$\overline{B}_{stock} = \sum Bcum_i = \sum \overline{B}_i \qquad (\overline{SB}_{stock} = \sum \overline{SB}_i)$$

$$\overline{W}_{stock} = \overline{B}_{stock} / \overline{N}_{stock}$$

$$C = \sum C_i$$

$$Y = \sum Y_i$$

$$\overline{W}_{catch} = Y / C$$

Several long–term projections can be made with different values of F_i, that is, with several fishing levels, $\overline{F}$, and/or with several exploitation patterns, s_i.

As mentioned before, the analyses of the effects of the fishing pattern on the catches and stocks can be done with the two components (fishing level and exploitable pattern), separated or combined.

Figure 4.4 (A-C) illustrates several types of yield per recruit curves, maintaining a fixed exploitation pattern. The curves are different for other exploitation patterns.

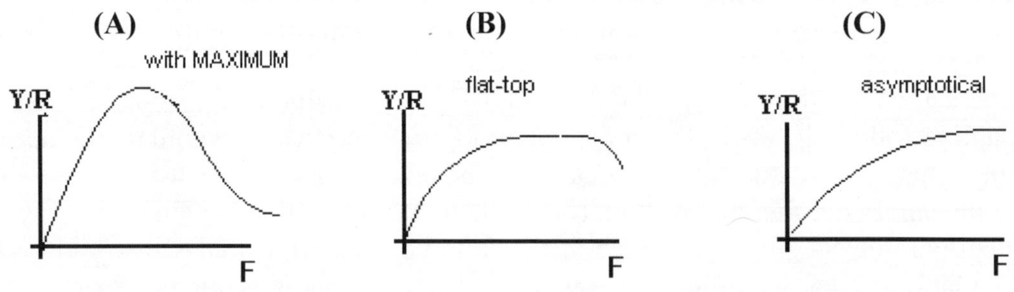

Figure 4.4 Examples of types of curves of the Yield per Recruit (Y/R) against F, given the *exploitation pattern*: (A – with a maximum, B - flat-top, C – asymptotic)

Figure 4.5 (A-E) illustrates the relations between the most important characteristics of the stock and fishing level, maintaining a stable exploitation pattern:

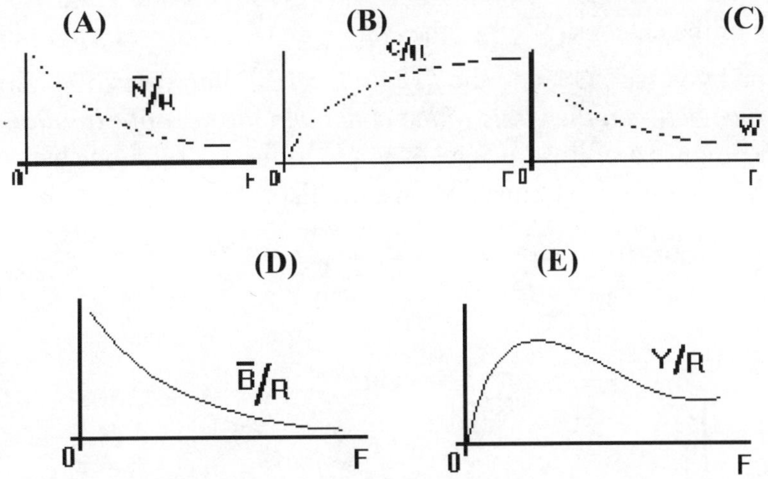

Figure 4.5 Relations between the long–term characteristics of the stock against the fishing level F, (A – mean number/R, B – catch in number/R, C - mean weight in the catch, D – mean biomass /R, E – yield/R)

A more detailed analyses of these curves will be presented later on, in Chapter 5.

Conclusion

The long–term structure of the stock by ages during 1 year	=	**Evolution of the cohort during its life**

Comments

1. One projects a cohort during its life in order to obtain the long–term projection of the stock for one year, assuming the annual recruitments to be constant.

2. It is necessary to know M_i for all ages of the cohort, as well as W_i , $\overline{W}$, s_i and the fishing level, $\overline{F}$, that is assumed to be constant in the years that follow.

3. Any recruitment size can be used. Adopt R = 1000 (or R = 1) with worksheets on your computer.

4. The five most important characteristics of the stock are $\overline{N}_{stock}$, $\overline{B}_{stock}$, C, Y and $\overline{W}_{catch}$ (see previous pages in this chapter for the respective expressions of calculation)

5. A characteristic of the stock that is also important is the spawning biomass, SB. To calculate SB, it is necessary to know the maturity ogive (or histogram).

6. Long–term projections are also designated as equilibrium situations.

7. Long–term projections are useful to define the long–term management objectives.

8. Annual $\overline{W}_{catch}$, are independent of the recruitment size (such as $\overline{L}_{catch}$ and $\bar{t}_{catch}$).

9. Economists transform the *total yield,* Y. into value Y$, *the mean weight of the catch,* $\overline{W}_{catch}$ into price of the catches, $\overline{W\$}_{catch}$, the *catch per vessel* (*or the cpue*) into value of the production by vessel, U$, and the *fishing level,* $\overline{F}$ into costs of exploitation, F$. The *difference between the value of the catch and the cost of exploitation,* Y$-F$, is the profit, L$. Figure 4.6 illustrates an example of the LT relations between those characteristics used by the economists against the fishing level, $\overline{F}$.

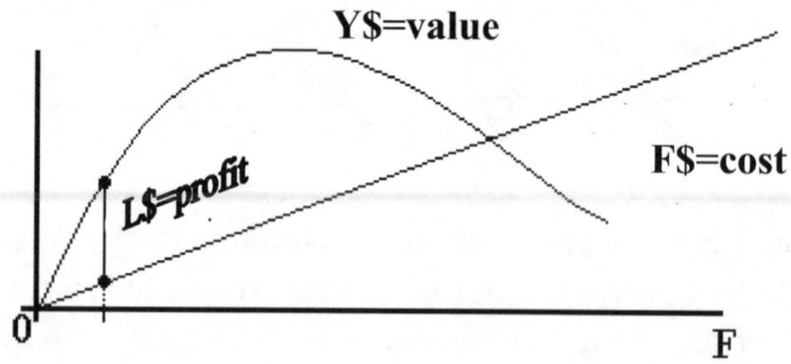

Figure 4.6 **Long–term relations between the value of the catch, the cost of exploitation and the profit against the fishing level**

4.5 STOCK–RECRUITMENT (S–R) RELATION

The stock–recruitment relation, known by S–R relation, associates the size of the stock, for one year, with *the recruitment which results* from the stock spawning normally during that year. The recruitment can be the recruitment at the exploitable phase and the stock can be the spawning stock. This recruitment may occur one or more years after the spawning.

The problem with the relation between the parental population and the new generation is not a special case of the fisheries resources, it is common to all the self renewable populations.

It is important to determine, in each case, the stock and the recruitment to be used. In fact, that *stock* can be the total number of individuals (at the beginning of the year or the mean value during the year), the number of adult individuals of the stock, the number of adult females, etc. One can also adopt, not the abundances in number, but the corresponding biomasses. The decision will depend on the type of resource and on the available data. It is necessary to define if the recruitment is in weight or in number and if the recruitment is to the fishing area or to the exploited phase.

In this manual, stock (S) will be considered as the spawning biomass and the recruitment (R) will be expressed in number.

After the spawning, the individuals of the new generation will have to go through different phases of the biological cycle: eggs, larvae and juveniles, before they become recruits. These phases, which, in some species can take years, are not directly submitted to the fisheries exploitation. That is why fishing in those years does not directly affect the size of the new recruitment. It is true that fishing acts on the size of the parental biomass, but that does not happen in the pre-recruit phases, and it is precisely for that interval of time that the relation S–R is applicable.

There is, in these pre-recruit phases, a great mortality due to climate and environmental factors (winds, currents, temperatures, etc.) as well as due to biological factors (available food, predation and others).

A great variety of factors (besides fishing) cause great fluctuations to the recruitment sizes, therefore, the relation S–R is a complex one. In conclusion, **the relation S–R, between the stock and the resulting recruitment**, can be considered **independent of fishing**.

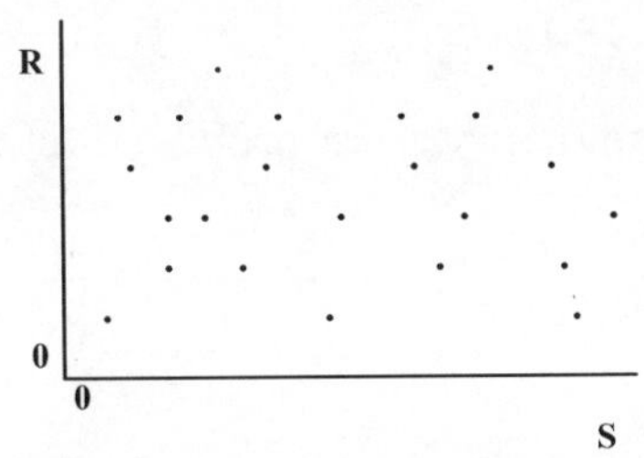

Figure 4.7 Example of the dispersion of the points in the relation S–R

Figure 4.7 shows the type of dispersion of values in the plot of recruitment (R) against parent stock (S).

Despite the difficulties, some models have been proposed for the relation S–R. One of the reasons for this is that there must be a relation. One point of the relation (S = 0, R = 0) is even known already.

4.5.1 BEVERTON AND HOLT MODEL (1957)

Beverton and Holt, when analysing plaice fishery in British waters proposed one model, which can be re-written as:

$$R = \frac{\alpha.S}{1+\frac{S}{k}}$$

where α and k are constants.

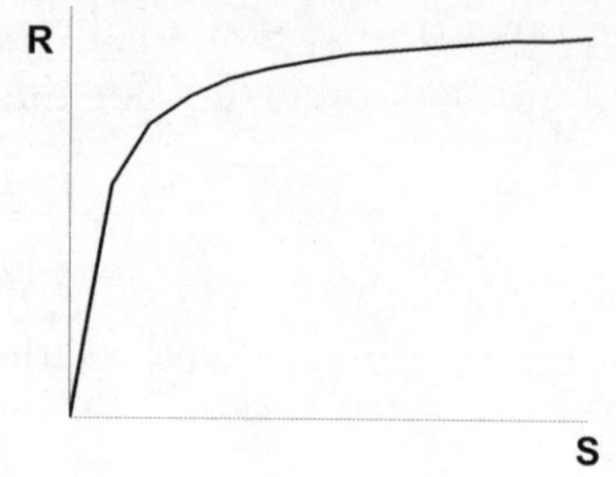

Figure 4.8 Beverton and Holt Relation S–R

This model shows a curve where R is asymptotic when $S \to \infty$.

4.5.2 RICKER MODEL (1954)

Ricker, studying cod fishery in Canadian waters proposed another model that can be written as:

$$R = \alpha.S.e^{-S/k}$$

where α and k are constants.

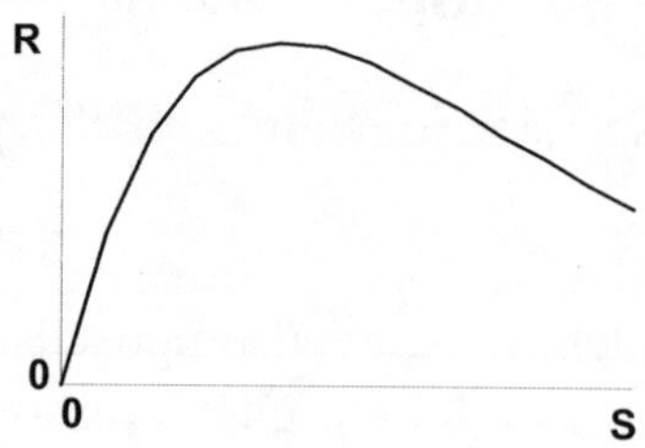

Figure 4.9 Ricker Relation S–R

This model presents a maximum resulting recruitment at an intermediate value of the parental stock, S.

4.5.3 OTHER MODELS

Other models of S–R have been proposed, like the **Deriso generalized model (1980)**, which can be re-written (Hilborn & Walters, 1992) as :

$$R = \alpha.S.(1 - c.\frac{S}{k})^{1/c}$$

(the model is the Beverton and Holt for c = -1 and when $c \to 0$ it is the Ricker model)

or **Shepherd's generalized model** (1982), as:

$$R = \frac{\alpha.S}{1+\left(\frac{S}{k}\right)^c}$$

(the model is the Beverton and Holt when c = 1 and when c = 2 the curve is an pproximation of the Ricker model)

Notice that in every model presented, S=0 implies R=0, as expected and the slope of the tangent to the curve at that point, (0,0), is equal to the parameter α. Sometimes, the relations S–R are presented as R/S in function of S, so the Beverton and Holt equation would be :

$$\frac{R}{S} = \frac{\alpha}{1+\frac{S}{k}}$$ showing that the inverse of R/S is linear with S.

With the Ricker model the relation between R/S against S would be:

$$\frac{R}{S} = \alpha.e^{-S/k}$$

that is, R/S is exponential negative with S (or ln(R/S) is linear with S).

With the Deriso model it would be $(R/S)^c$ linear with S.

Mathematically, these linear relations can be useful to estimate the parameters α and k, but statistically they cause some problems, because the variable S appears in the response variable and in the auxiliary variable.

Comments

1. Remember that relations S–R are independent of the fishing level.
2. Relations S–R may be introduced in the calculations of the stock projections. In that case, the projections will have to be made every year and they will require the structure of the stock at the beginning of the initial year.
3. It must not be forgotten that there is a great dispersion of the observed points (S, R) around the model.
4. To determine the limit of the Deriso equation when $c \to 0$ it is enough to remember that limit $(1+A/n)^n = e^A$ when $n \to \infty$ and A is a constant.
5. The Beverton and Holt model presents an asymptote, $R \to \alpha.k$ when $S \to \infty$. When S=k it will be $R = \alpha\ k/2$.

The Deriso model presents the maximum:

$$S_{max} = k / (1+c),\ R_{max} = \alpha.k / (1+c)^{1+(1/c)}$$

The Shepherd model presents the maximum:

$$S_{max} = k.(c\text{-}1)^{-1/c},\ R_{max} = (\alpha /c).k.(c\text{-}1)^{1-(1/c)}$$

4.6 RELATION BETWEEN R AND $\overline{B}$ (R–S RELATION)

Up to now the discussion have been centred on the relation S–R, that is the relation between the biomass S and the resulting recruitment R. There is another relation (which has already been referred to in Section 4.4 Long–term projections of the stock, particularly in the conclusion about the structure of the stock and the evolution of a cohort during all its life) that could be called the relation R–S, that is the relation between the recruitment R and the **resulting cumulative biomass**, B_{cum} of a cohort during all its life for a given fishing pattern.

The cumulative biomass (or spawning biomass) of a cohort during its life, B_{cum} , is, as it has already been seen, equal to:

$$B_{cum} = R\ .\ \{\text{function of biological parameters and of the fishing pattern, F}\}$$

As mentioned before, the cumulative biomass, B_{cum} , of a cohort during its life is equal to the long–term *mean biomass*, $\overline{B}$, of the stock during one year, so, one can refer to the B_{cum} as $\overline{B}$.

It can then be said that in the long–term, if the biological parameters are considered constant, then for a certain fishing pattern, the mean biomass of a stock during one year is proportional to the recruitment. Figure 4.10 illustrates the proportionality.

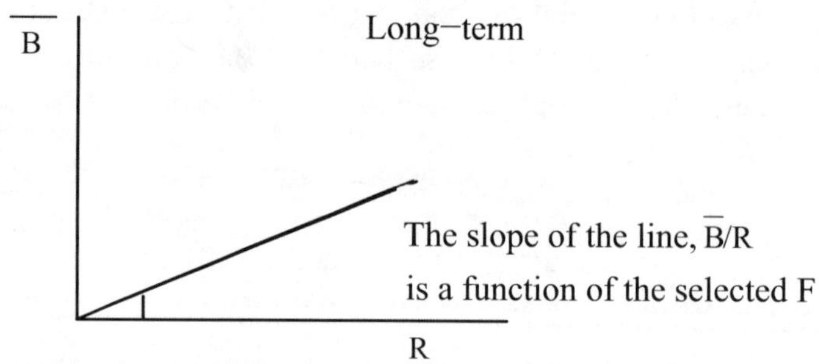

Figure 4.10 Illustration of the proportionality in the long–term between the mean Biomass $\overline{B}$ of a stock and the Recruitment (R), for a certain fishing level, F, assuming that the exploitation pattern and the biological parameters are constant

Notice that in Figure 4.5-D, the relation B/R against F was shown. The two curves 4.5-D and that of Figure 4.10 are different representations of the same situation. While in Figure 4.5-D, a value of F corresponded to a value of B/R given by the curve, in Figure 4.10 the value of F will correspond to a slope B/R of a given straight line.

For other values of F, the straight line of Figure 4.10 will have different inclinations. Figure 4.11 shows several lines according to the different values of F.

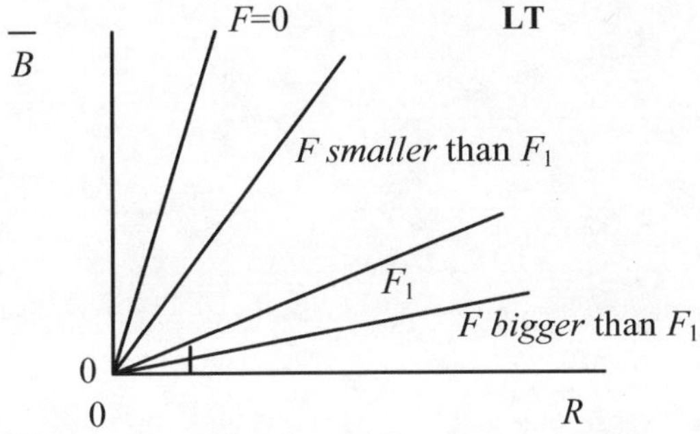

Figure 4.11 Illustration of four lines corresponding to different fishing levels, F

Figure 4.12 shows the overlapping of the Ricker curve S–R (Figure 4.9) and the line in Figure 4.10, the axes of this last figure having been switched. To avoid confusions, it is convenient to refer the slope of the line corresponding to a value F in relation to the axis R.

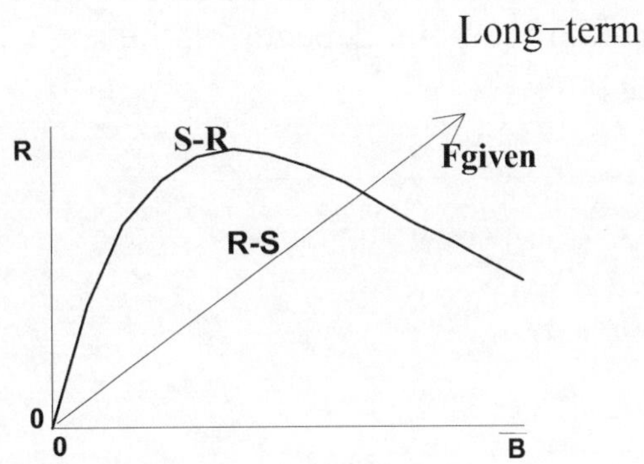

Figure 4.12 Overlapping of the curve S–R with the line R-S for a given F

One can start from a value of biomass, and, through the relation S–R, determine the future recruitment, R. That R will give a resulting biomass, $\overline{B}$, through the straight line. This process can be repeated until a situation of equilibrium is found.

It will be possible to illustrate the combined analysis through the two relations for a certain fishing level.

For example, let us select one value $\overline{B}_1$. The curve S–R allows the calculation of the resulting value R_2. Through the line (of a given F), the resulting value of $\overline{B}_2$ will correspond to that value of R_2, and so on. Figure 4.13 shows that the process reaches the equilibrium point (R_E, $\overline{B}_E$). This point would correspond to a value of the fishing level F, determined by the slope of the line in relation to the R axis.

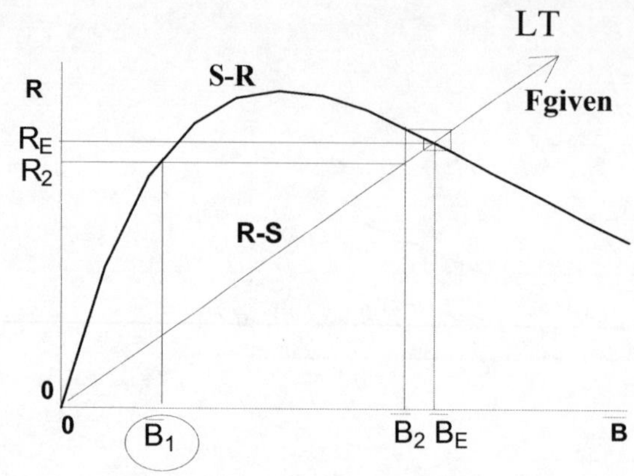

Figure 4.13 Illustration of the process that, starting from a biomass $\overline{B}_1$, theoretically leads to the equilibrium point (R_E, B_E)

Notice that the intersection of the two relations does not always lead to an equilibrium point. The existence of an equilibrium point depends on the angle between the straight line and the curve at the intersection point.

CHAPTER 5 – BIOLOGICAL REFERENCE POINTS AND REGULATION MEASURES

5.1 BIOLOGICAL REFERENCE POINTS FOR THE MANAGEMENT AND CONSERVATION OF FISHERIES RESOURCES

The long–term objectives for fisheries management should take into consideration scientific fishing research and population dynamics, as well as the climatic changes that may affect the stocks.

In order to define these long–term objectives we have to consider the values of the fishing level, which allow bigger catches in weight, whilst also ensuring the conservation of the stocks. The extreme values of the biomass or the fishing level, which might seriously affect the self renovation of the stocks, also have to be considered. These fishing level values, of catch and biomass are designated as biological reference points (BRP). In this manual some of the different types of BRP will be considered (Caddy, & Mahon, 1995; FAO, 1996 and ICES, 1998).

The Target Reference Points, TRP are BRP defined as the level of fishing mortality or of the biomass, which permit a long–term sustainable exploitation of the stocks, with the *best possible catch*. For this reason, these points are also designated as *Reference Points for Management*. They can be characterized as the *fishing level* F_{target} (or by the *Biomass*, B_{target}).

The most well known F_{target} is $F_{0.1}$ but other values, like F_{max}, F_{med}, and F_{MSY} will also be studied.

For practical purposes of management, the TRP will be converted, directly or indirectly, into values of fishing effort, given as percentages of those verified in recent years.

The Limit Reference Points, LRP are maximum values of fishing mortality or minimum values of the biomass, which must not be exceeded. Otherwise, it is considered that it might endanger the capacity of self-renewal of the stock.

In the cases where fishing is already too intensive, the LRP can be important to correct the situation or to prevent it from getting worse.

The LRP are limit values, mainly concerned with the conservation of marine stocks and they are therefore also referred to as *reference points for conservation* (this designation does not imply that the F_{target} are not concerned with conservation).

Several LRP have been suggested, which will generally be referred to as F_{lim} or B_{lim}. In this manual the levels of biomass B_{loss} and MBAL will be referred to as B_{lim} and the fishing levels F_{loss} and F_{crash} as F_{lim} .

The Precautionary Principle, proposed by FAO in the Conduct Code for Responsible Fisheries (FAO, 1995), declares that the limitations, uncertainties or lack of data for the assessment or for the estimation of parameters, cannot be justification for not applying regulation measures, especially when there is information that the stocks are over-exploited.

From this point of view, it is important to make clear which basic assumptions are necessary in order to estimate the consequences on the catches and on the abundance of the stocks.

The uncertainties associated with the estimation of F_{lim} , and B_{lim} , therefore lead us to determine new reference points, called Precautionary Reference Points, Fpa or Bpa.

The assumptions and the consequences of adopting alternative hypotheses about the stock and fishing characteristics should always be presented to justify the estimated values of Fpa (or Bpa).

The new limits (Fpa or Bpa) due to the application of the Precautionary Principle, will be more restrictive than the LRP′s. The practical consequences of these new limits are the regulation measures designed to control the fishing effort which are more severe than in those cases where there is appropriate data.

It can be said that this is the price to pay for not having the appropriate conditions to make available reliable data and information.

The Precautionary Approach, suggests that the results of fisheries research should be adopted by management with regard to the formulation of the regulation measures and that these measures should also take into consideration the *socio-economic and technical conditions of fishing* (FAO, 1996).

A final remark about all the Biological Reference Points mentioned above :

The evaluation of the biological reference points has to be updated, taking into consideration the possible changes in the biological parameters or any other necessary correction of the exploitation pattern. This fact is important because the new biological reference points will be different from the previous ones.

5.2 BIOLOGICAL TARGET REFERENCE POINTS (F_{max}, $F_{0.1}$ F_{med} and F_{MSY})

5.2.1 F_{max}

Definition

1. Consider the long–term yield per recruit, Y/R, as a function of F, for a certain exploitation pattern.

   F_{max} is the point of the curve, Y/R against F, where Y/R is maximum.

   Figure 5.1 shows a curve of Y/R against F.

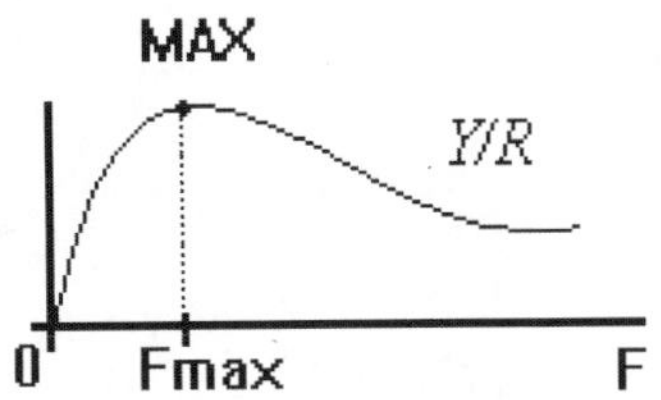

Figure 5.1 **Y/R as function of F for a certain t_c constant, showing F_{max} and Y_{max}**

In Chapter 4 it was mentioned that to estimate the long–term projections one could assume that the recruitment is constant and equal to 1 (R=1). In this way, the mathematical expressions are sometimes written with Y instead of Y/R.

2. Mathematically, at point F_{max}, the derivative of Y/R against F is equal to zero, that is,

For $F = F_{max}$ will be $\frac{dY}{dF} = 0$ air$(Y) = 0$ (Value of Y is maximum)

For $F < F_{max}$ will be $\frac{dY}{dF} > 0$ air$(Y) > 0$ (Y is increasing with F)

For $F > F_{max}$ will be $\frac{dY}{dF} < 0$ air$(Y) < 0$ (Y is decreasing with F)

Geometrically, the slope of the tangent to the curve is equal to zero for $F = F_{max}$, positive for $F < F_{max}$ and negative for $F > F_{max}$.

Comments

1. $\overline{B}_{max}/R$ and Y_{max}/R are the values at F_{max}.

   It is also convenient to analyse the situation of $\overline{B}/R$ at the points $F \neq F_{max}$.

   $F < F_{max}$ corresponds to $\overline{B} > \overline{B}_{max}$

   $F > F_{max}$ corresponds to $\overline{B} < \overline{B}_{max}$

   Point F_{max} does not depend on the value of the recruitment.

2. For another relative pattern of exploitation there will be another F_{max}.

3. All the points of the curve Y/R against F, are long–term points or equilibrium points.

4. When the level, F, is bigger than F_{max} it is said that there is growth overfishing.

   It is convenient to present the two curves Y/R and $\overline{B}/R$ against F, in the same graph (usually with different scales).

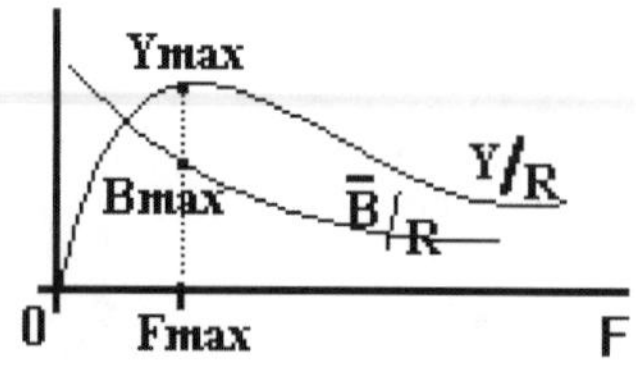

Figure 5.2 Long–term curves of Y/R and $\overline{B}$/R against F, given an exploitation pattern

F_{max} was adopted by the majority of the International Fisheries Commissions as a long–term objective of management (1950-1970).

Even today F_{max} is used as a target-point having been proposed as a Limit Reference Point (LRP) in some cases.

The flat-top and asymptotical curves do not allow the determination of an F_{max}.

The definition of F_{max} does not consider the appropriate level of spawning biomass.

F_{max} only indicates the value of F which gives the maximum possible yield per recruit from a cohort during its life, for a given exploitation pattern.

The analyses of these long–term curves, mainly of $\overline{B}$, Y and $\overline{W}_{catch}$ against the *fishing level*, give information about the abundance of the resource (or catch per vessel), total yield of all the fleet and mean catch weight for different fishing levels.

5.2.2 $F_{0.1}$

1. ***Definition***

Consider the long–term yield per recruit, Y/R, as a function of the coefficient of fishing mortality, F. One value of air(Y/R), corresponds to each fishing level, F. The air(Y/R) is maximum when F = 0 and decreases, being zero when $F = F_{max}$.

The point $F_{0.1}$ is the value of F where air(Y/R) is equal to 10 percent of air(Y/R) maximum.
The figure 5.3 illustrates this situation.

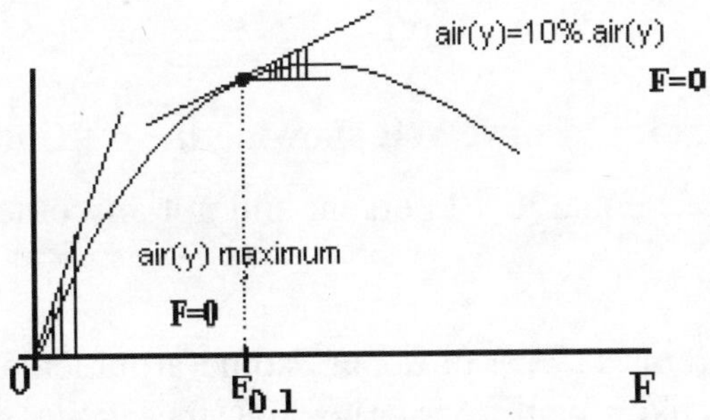

Figure 5.3 Y/R showing the reference target point $F_{0.1}$

2. For F = 0, the biomass per recruit, $\overline{B}$/R will be $\overline{B}_0$/R, also designated as *Virgin Biomass* or *Non-exploited Biomass*. The air(y) at F=0 is also equal to B_0

In fact, $Y = F \cdot \overline{B}$ implies $\frac{dY}{dF} = \overline{B} + F\frac{d\overline{B}}{dF}$

Then, for $F = 0$, $\text{air}(Y) = \left(\frac{dY}{dF}\right)_{F=0} = \overline{B}_0$

So, from the definition given in point 1 one can also say that $F_{0.1}$ is the value of F where air(Y) = 10 percent of the virgin biomass.

3. ***Calculation of* $F_{0.1}$**

Let the function $V = Y - 0.1.\overline{B}_0.F$

It can be proved that the function V is maximum when $F = F_{0.1}$

In fact, V is maximum when $\frac{dV}{dF} = 0$, then:

$$\frac{dV}{dF} = \frac{dY}{dF} - 0.1 \cdot \overline{B}_0 = 0 \quad \text{or} \quad \frac{dY}{dF} = 0.1 \cdot \overline{B}_0$$

Therefore, the value of F corresponding to the previous dY/dF is the value of $F_{0.1}$.
$F_{0.1}$ can then be calculated by maximizing the function $V = Y - 0.1.\overline{B}_0.F$

$\overline{B}_0$ can be calculated, for example, from the long–term relation of $\overline{B}$ against F, when F = 0.

Graphically it will be:

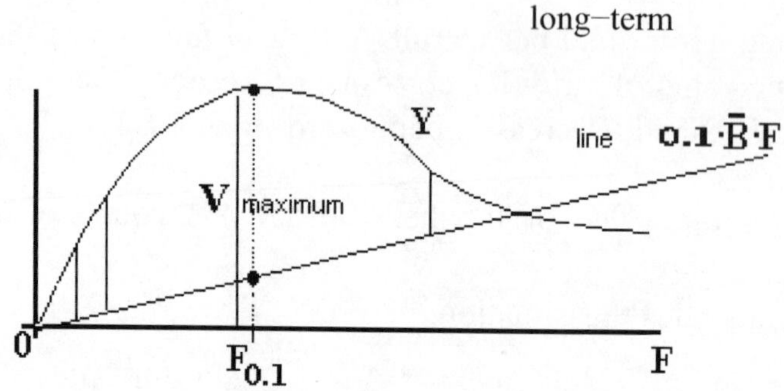

Figure 5.4 **Curve Y/R showing the maximum of the function V**

4. Why adopt air(Y/R) equal to 10 percent and not any other percentual value, for example, 20 percent?

Gulland and Boerema (1969) presented some arguments, including financial arguments. Some countries (like South Africa) adopt the value of 20 percent with a resulting reference point $F_{0.2}$ that is more restrictive than $F_{0.1}$.

5. Figure 5.5 illustrates the two biological reference points F_{max} and $F_{0.1}$.

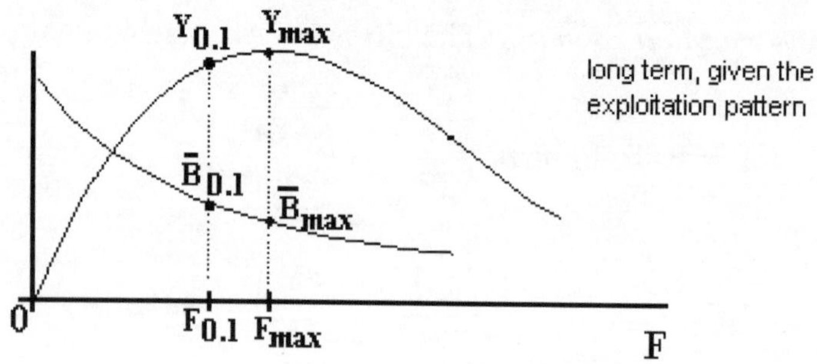

Figure 5.5 **Long–term variation of Y/R and B/R against F and points corresponding to F_{max} and $F_{0.1}$**

$\overline{B}_{0.1}$ and $Y_{0.1}$ are the values of $\overline{B}$ and Y corresponding to $F_{0.1}$

$F_{0.1}$ is always smaller than F_{max}

$\overline{B}_{0.1}$ is always larger than $\overline{B}$

$Y_{0.1}$ is always smaller than Y_{max}, although, in practice, the difference is not large.

The second sentence above indicates the advantages of $\overline{B}_{0.1}$ over B_{max}. The last sentence shows that $Y_{0.1}$ is not the largest possible catch, but is acceptable as a target point of management. The fact that $\overline{B}_{0.1}$ is larger than $\overline{B}_{max}$ suggests that the fishing level $F_{0.1}$ is preferable to F_{max} as TRP.

Notice that $F_{0.1}$ can be calculated even when the curve is asymptotical or flat-top.

Another value of $F_{0.1}$ will be obtained if the exploitation pattern changes.

In the years 1960-70 $F_{0.1}$ started to be preferred to F_{max} as a target point by resource managers, and it was adopted in the 80's, as a long–term objective by many International Fisheries Commissions and by the EEC.

5.2.3 F_{med}

1. ***Definition***

This target point considers the relation S-R between the stock and the resulting recruitment, in order to avoid the assumption of a constant recruitment.

Let (the spawning or total) biomasses and the resulting recruitments for each year of a certain period of time be known. In this case, one can calculate the median value of the ratios between the annual spawning biomasses and the corresponding recruitments.

F_{med} is defined as the F value corresponding to the median B/R ratio in the long–term B/R relation against F.

Usually, F_{med} is illustrated by considering the graph of the points corresponding to pairs of values of parental biomass (total or spawning), $\overline{B}$, during that year and the resulting recruitment , R. Figure 5.6 shows this situation.

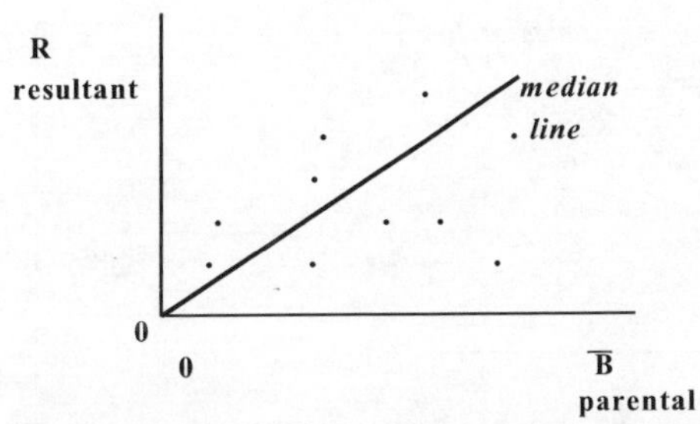

Figure 5.6 Illustration of a median line

The marked line is a line passing through the origin, which separates the total number of points in equal parts, that is, 50 percent of the points are in the upper part and 50 percent are in the lower part of the line. This line is designated as the median line, or 50 percent line, which can be explained as follows: in 50 percent of the years of the considered period the values of R were smaller than the values of R which were estimated by the median line (or, in 50 percent of the years of the referred period the values of R were bigger than the values of R estimated by the median line).

As seen in Section 4.5 the slope (B/R) of each line marked in the graph, is associated with a certain value of the fishing level, F. The value of F associated with the median line is then, the median target point, F_{med}.

It can be said that, given a certain level of parental biomass and knowing the $\overline{B}/R$ corresponding to F_{med}, then there is a 50 percent probability that the resulting recruitment will be less than (or greater than) the value indicated by the median line.

2. *Calculation of* F_{med}

In order to determine the value of F_{med} it is necessary to consider the long–term relation between the resulting biomass per recruit and the fishing level, F, (Section 4.4, Figure 4-D).

The determination of F_{med} can be done mathematically or graphically.

To make the mathematical calculation of F_{med} the ratio $\overline{B}/R$ has to be determined for each pair of values $(\overline{B}, R)$. Those values have to be ordered and then the median value, $\overline{B}/R_{median}$ can be calculated.

In the long–term relation of the Biomass per Recruit against F, the value of F_{med} is the value of F that corresponds to the median value previously obtained.

To make the graphical calculation, notice that in Figure 5.6 the slope of the median line against axis R is equal to $\overline{B}/R$. This value will be the basis for the calculation of value F_{med} in the graph of $\overline{B}/R$ against F, in the long–term projection. Figure 5.7(A-B) illustrates the calculation.

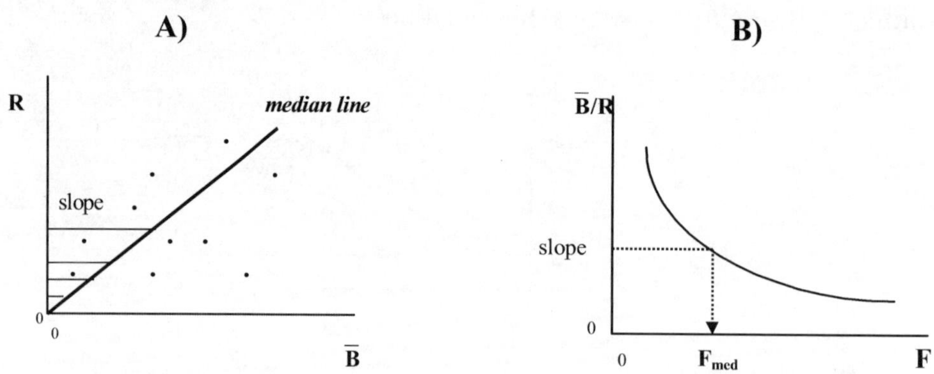

Figure 5.7 Illustration of the graphical calculation of F_{med}

Other straight lines, corresponding to probabilities other than 50 percent, can be marked. Figure 5.8 illustrates the graphical calculation of $F_{10\%}$. The line marked on Figure 5.8A, separates the points, in such a way that 10 percent of them remain below the line (or 90 percent of the points stay above the line). So, this line has been designated as the 10 percent line.

Notice that the slope of the 10 percent line with axis R is larger than the slope of the median line, as shown in Figure 5.9B. In this way, $F_{10\%}$, or F_{low}, is smaller than F_{med}.

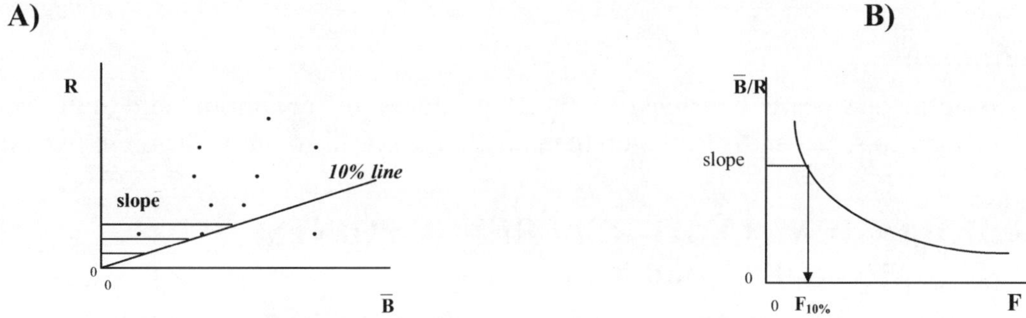

Figure 5.8 **Illustration of the graphical calculation of $F_{10\%}$**

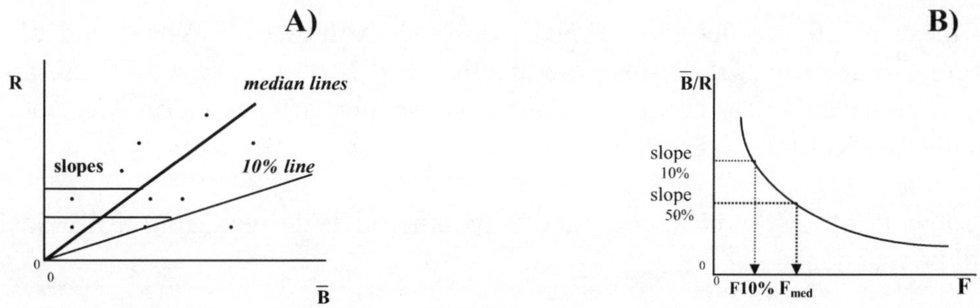

Figure 5.9 **Graphical illustration of the slopes and corresponding F´s for the median and 10 percent lines**

Comments

1. The target point F_{med} intends to ensure an acceptable level of biomass based on the empirical relation S-R.
2. Other percentages can also be adopted, corresponding to straight lines which estimate different probabilities of recruitments which are less than those indicated by the median line. So, F_{high} would be the fishing level corresponding to the 90 percent line, for which the recruitments of 90 percent of the observed years would be less than those estimated by the line.
3. $F_{90\%}$, as will be seen in the following Section, is also considered as a Limit Point (LRP).
4. Notice that the slope of the 10 percent line with axis R is larger than the slope of the median line and therefore, $F_{10\%}$ (or F_{low}) is smaller than F_{med} (Figure 5.9B).
5. F_{med} was used in management, in recent years, particularly with Iberic sardines.
6. The biomass used can be the total biomass, $\overline{B}$, but is frequently the spawning biomass, SP.
7. If the median line does not pass through a marked point in the scatter plot (which happens when there is an even number of points) then one can use any straight line passing between the two central points, for instance, the mid-line. In any case, F_{med} is always an approximate value.

5.2.4 F_{MSY}

Definition

F_{MSY} is defined as being the value of F which produces the maximum yield in the long–term. It is necessary to select an S-R relation to estimate F_{MSY}. This point is different from F_{max}.

5.3 BIOLOGICAL LIMIT REFERENCE POINTS (B_{loss}, MBAL, F_{crash} and F_{loss})

There are several proposals for F_{lim} and B_{lim}. For each stock, the adopted values of F_{lim} and B_{lim} depend on the characteristics of the stock and on its exploitation. What is important is that the adopted LRP be a value that allows an exploitation level which avoids dangerous situations of stock renewal.

Some of these points are derived from the observed values of Biomass and of Resulting Recruitment. Some examples of this type are B_{loss} and MBAL. These LRP are also usually classified by some authors as non-parametric, because their determination does not depend on any particular model of the S-R relation.

Another category of LRP points, classified as parametric, is derived from S-R models. F_{crash} will be mentioned.

Let us also mention the category of LRP points involving observed values and values obtained by the application of S-R models, like, for example, F_{loss}.

5.3.1 B_{loss}

B_{loss} is the smallest spawning biomass *observed* in the series of annual values of the spawning biomass (Lowest Observed Spawning Stock).

5.3.2 MBAL

More satisfactory is the LRP designated as Minimum Biological Acceptable Level, MBAL. In fact, this LRP is a spawning biomass level below which, observed spawning biomasses over a period of years, are considered unsatisfactory and the associated recruitments are smaller than the mean or median recruitment.

5.3.3 F_{crash}

The name itself shows that it is a limit that corresponds to a very high value of F, showing a great probability of collapse of the fishery.

F_{crash} is the fishing level F which will produce a long–term spawning biomass per recruit (S/R) equal to the inverse of the instantaneous rate of variation of R with the biomass, at the initial point (S = 0, R = 0). With the S-R models of Section 4.5 that value is the parameter $1/\alpha$ of the models.

In order to make the graphical determination of this LRP one can start by obtaining the slope of the angle that the tangent to the S-R curve makes with the R axis at the origin. Afterwards, and starting from the relation $\overline{B}/R$ against F, the value of F that corresponds to the value

$\overline{B}/R$ indicated by that slope is obtained. F_{crash} will then be the value of F corresponding to $\overline{B}/R$ equal to that slope, in the long-term relation $\overline{B}/R$ against F.

5.3.4 F_{LOSS}

F_{loss} is usually defined as the fishing level F which will produce a long–term spawning biomass per recruit (S/R) associated to B_{loss}.

To determine this limit point, first obtain the value of R corresponding to B_{loss} on the adjusted curve S-R. Then, calculate B_{loss}/R and find the value of F, in the long–term relation B/R against F.

Most of the Limit Points shown have been criticised for depending on the observed values or on the adjustment of the S-R relation.

5.4 PRECAUTIONARY REFERENCE POINTS - Fpa, Bpa

As previously mentioned, the Precautionary Principle recommends that the assessments should be done even when the basic data presents some gaps. This recommendation implies that, in this case, the determination of the Biological Reference Points will not be very precise. The uncertainities of the estimates should be calculated, and it is necessary to mention the assumptions and models which have been used.

One suggestion to determine Fpa and Bpa might be to estimate F_{lim} or B_{lim} and from these values, to apply the following empirical rules :

$$Fpa = F_{lim}.e^{-1.645.\sigma} \quad \text{and} \quad Bpa = B_{lim}.e^{+1.645.\sigma}$$

The constant σ is one measure associated with the uncertainity in the estimation of the fishing mortality level, F. The values obtained in several fisheries indicate that values of σ are within the interval (0.2, 0.3) (ICES, 1997). In practice, it can be said that Fpa is between $0.47F_{lim}$ and $0.61F_{lim}$,and Bpa is between $1.39B_{lim}$ and $1.64B_{lim}$.

It is important to make clear that the target points may also, in certain cases, be considered as limit or precautionary points depending on the combined analyses of the exploitation of the stock and of the biological reference points obtained.

5.5 FISHERIES REGULATION MEASURES

The regulation measures aim to control the fishing level and the exploitation pattern applied to the stock for an adequate exploitation.

The most common regulation measures to control fishing levels are:

- Limitation of the number of fishing licences.
- Limitation of the total fishing effort each year (limiting fishing days, number of trips, number of days at sea, etc.).
- Limitation of Total Allowable Catch (TAC)

TAC is a measure that directly controls the catch and, indirectly, the fishing level. It is convenient to combine the TAC with the allocation of quotas of this total TAC for each component of the fleet. In this way, the competition between vessels to fish the maximum possible catch, as quickly as possible, before the TAC is reached, can be avoided.

The system of quotas allocated to each vessel is called Individual Quotas (IQ).

The regulation measures to correct the exploitation pattern are usually called technical measures. Some of these measures are :

- The minimum size (or weight) of the landed individuals.
- The minimum mesh size of the fishing nets.
- The prohibition of fishing in spawning.
- The closed areas and periods for the protection of juveniles.

The fishing management have the duty to promote legislation and the application of the regulation measures. (In the particular case of the EU, and for the stocks of the Economic Exclusive Zones (EEZ's) of the member states, the Commission decides on the measures to be taken). In any case, management needs the analyses on the state of the stock and its exploitation and on the effects of the recommended measures. That study must be done by the fisheries scientists of each country or region and their Fishing Research Institutes, who will have to calculate the projections of the stocks. The International Council for the Exploitation of the Sea (ICES) analyses the assessments and recommends regulation measures and the expected effects of the application of those measures to the Commission, as well as the consequences of their non-application.

The short–term projections, as well as the regulation measures, only make sense if the long–term objectives of fisheries management are previously analysed and defined. Short–term projections of the stock and of its fishing must also be made by the scientists.

Comments

1. Management needs to define the fishing objectives, based on the long–term projections. Those objectives are valid for a period of years, even if they can be adjusted every year.

2. The regulation measures, on the contrary, have to be established every year, although some of them may be valid for more than one year. Some technical measures, like the minimum mesh sizes of the fishing nets or the minimum size of the landed individuals, are valid for several years.

3. All the measures have advantages, difficulties and disadvantages regarding the purposes they intend to reach.

   The concession of fishing licences is a common practice almost everywhere, with a limited total number.

   TACs and quotas, because they control the catches, have caused misleading declarations about catches.

   The direct limitation of the total fishing effort (f), is based on the assumption that the measure causes a similar limitation on the fishing mortality coefficient (F). However, this relation may not be proportional. In the first place, it is difficult to measure the fishing effort of the different fishing gears and of all the involved fleets and it is also difficult to express it in units that respect the proportionality between F and f. Secondly, the capturability of several gears may increase (and consequently increase F*)* without increasing the fishing effort. Finally, the expected proportionality between F and f may not be true. In any case, what matters, is not to forget that there is a relation between F and f.

4. The protection of the juveniles should be carried out during the whole year and will preferably control the fishing mortality throughout the year. The occasional measures, like areas and periods when fishing is forbidden for the protection of juveniles, require annual investigations in order to discover whether there are exclusive concentrations of juveniles, to assess the effects of that occasional interdiction, and to find out the consequences of the interdiction on other species, etc. The minimum size of the landed individuals, does not mean that smaller individuals are not caught, but only that they are not landed. The difference between the catch and the landing is the so-called rejections to the sea. It is clear that if the individuals are caught and rejected to the sea, the fishing mortality is larger than the one suggested by the landings. The minimum landing size of the individuals may have the effect of dissuading fishermen from catching small individuals. Currently, some countries are forcing the landing of all fish caught.

5. The closed spawning areas and seasons, to save the spawning biomass and indirectly protect recruitment, is far from effective in the latter objective. In fact, large spawning biomasses correspond to a large number of eggs, but that does not necessarily imply bigger recruitments, as seen in Section 4.5. It is also not always true that forbidding fishing during the spawning, and not forbidding before (or after) the spawning, protects the spawning biomass. The only way to protect the spawning biomass will be to control fishing level during the whole year. Finally, it has to be said that, in any case, the interdiction of fishing in the spawning area and period, or on any other occasion, always represents a reduction of the fishing effort. This is not a major inconvenience and in some cases, may even be beneficial.

6. It has to be stressed that no regulation measure will accomplish its objectives without observing two conditions :

   The understanding of the fishermen (broadly speaking) that the measure is good for the fishery. Hence, it is important to discuss the scientists' conclusions, their objectives, their reasons and the expected effects.

An efficient fiscalization in the ports and at sea! The 200-mile exclusive zone may be very vast and the fiscalization very expensive, but it is not necessary to fiscalize the whole area intensively. It is enough to control the areas of larger catches more intensively and the remaining areas less so.

During the last few years, new ways of controlling access to fisheries resources and exploitation levels are being implemented. Some examples are the establishment of Individual Transferable Quotas systems (ITQ), co-management systems or even the system of regional or municipal management, where some management responsibilities are attributed to the resource users themselves.

7. The ITQ management system is based on the abusive assumption that only the economically efficient and profitable vessels deserve to be active in fishing. So, TAC's are divided into individual quotas, to be auctioned for the best offer.

The co-management system delegates a great part of the responsibility of management to those who directly exploit the fishing resources - managers, fishermen and their professional organizations or unions. With this system neither are quotas sold in auctions, nor are the fishing licences lost.

These are the most well known systems.

The ITQ system presents the following inconveniences : permanent loss of the titles of quotas and of fishing licences; concentration of quotas in the hands of a small group of people (who may not even belong to the fishing sector or are even foreigners); and underestimation of the social, human and cultural aspects, in favour of economic efficiency criteria.

On the contrary, the co-management system, is concerned with the social aspects of the people involved; it seeks their direct and conscientious co-participation with government authorities in the management responsibilities, including fiscalization.

CHAPTER 6 – PRODUCTION MODELS

6.1 BASIC ASSUMPTION ABOUT THE EVOLUTION OF THE BIOMASS OF A NON EXPLOITED STOCK

The production models (also called general production models, global models, sintetic models or Lotka-Volterra models) consider the stock globally, that is, they do not take into consideration the structure of the stock by age or size.

The total biomass of a *non exploited* stock cannot grow beyond a certain limit. The value of that limit depends, for each resource, on the available space, on the feeding facilities, on the competition with other species, etc. In conclusion, it depends on the capacities of the ecosystem to maintain the stock. That size limit of the biomass will be designated by *Carrying Capacity,* k.

The total biomass of a non exploited fishery resource has the tendency to increase with the time towards its carrying capacity, k, with a non constant absolute rate, air(B_t). The rate, air(B_t), is small when the biomass is small, increases when the biomass grows and is again small when the biomass gets close to the carrying capacity. Changes, including reductions, can occur in the biomass due to fluctuations of the natural factors, but, in any case, the tendency will always be an increase towards its carrying capacity.

The instantaneous rates air(B_t) or rir(B_t) are therefore not constant.

In order to formulate the basic assumption of a model for the evolution of the non-exploited biomass, one can adopt a function H of B_t , as was done with the basic assumption of the individual growth, and define it:

$$\underset{\text{natural}}{\text{rir}}[H(k)-H(B_t)] = -r \qquad \text{with r constant}$$

r is the intrinsic growth rate of B_t. The relative instantaneous rate, rir(B_t), of the non-exploited biomass can therefore be deduced as.

$$\underset{\text{natural}}{\text{rir}(B_t)} = \frac{r[H(k)-H(B_t)]}{\left[B.\dfrac{dH}{dB_t}\right]}$$

6.2 EXPLOITED STOCK

When the stock is exploited, the rate of variation of the biomass due to all causes, that is, the total rir(B_t), can be separated into two components: natural rir(B_t) due to all causes but not fishing and rir(B_t) due to fishing:

$$\text{rir}(B_t)_{\text{total}} = \text{rir}(B_t)_{\text{natural}} + \text{rir}(B_t)_{\text{fishing}}$$

In an interval of time T_i and with a constant fishing level, it will be:

$$rir(B_t)_{fishing} = F_i = \text{constant}$$

and:

$$rir(B_t)_{total} = f(B_t)_{natural} - F_{ifishing}$$

The natural rate, $rir(B_t)$, (which, according to the basic assumption of the natural evolution of the biomass, B, is supposed to be a function of the biomass, B_t) is usually designated as $f(B_t)$.

Comments

1. Historically, the production models were the first to be used on the analyses of the evolution of biological populations, Lotka-Volterra (1925-1928).
2. Schaefer (1954) applied a production model to a fish stock subject to fishing.
3. The carrying capacity, k , has been designated in fisheries biology, as B_∞ and also as B_0. Currently the symbol k is preferred (notice that this symbol is different from the symbols, K, of the individual growth models and of the relation S-R).
4. The basic assumption about natural $rir(B_t)$ previously presented, can be mathematically formulated in different ways..
5. The production models can only be used in fisheries to analyse the effects of fishing level changes and not of changes in the exploitation pattern, because the models consider the biomass in a global way and do not take into consideration the age or size stock structure.

6.3 VARIATION OF THE BIOMASS IN THE INTERVAL T_i

The "total", "natural" and "by fishing" instantaneous rates can be approximated by the relative mean rates, $rmr(B_t)$. In fact, it can be said that $rmr(B_t) \cong rir(B_t)$ relative to the mean $\bar{B}_i$. (This relation is exact in the case of the exponential model) .

The following general expression in terms of instantaneous rates

$$rir(B_t)_{total} = f(B_t)_{natural} - F_{ifishing}$$

can then be approximated, replacing the rates by the respective mean rates in relation to the mean biomass during the interval T_i:

$$rmr(B_t) = f(\bar{B}_i) - F_i \qquad \text{in relation to } \bar{B}_i$$

or

$$\frac{1}{\bar{B}_{i\ total}} \cdot \frac{\Delta B_i}{T_i} \cong f(\bar{B}_i) - F_{i\,fishing}$$

and

$$\Delta B_i \cong f(\bar{B}_i).\bar{B}_i.T_i - F_i.\bar{B}_i.T_i$$

The variation of the biomass due to all causes of mortality is then decomposed into the variation due to natural mortality and the variaton due to the fishing mortality:

$$\Delta B_{i\,total} \cong f(\bar{B}_i).\bar{B}_i.T_i - Y_{i\,fishing}$$

The value of the biomass, B_{i+1}, at the end of the interval T_i is:

$$B_{i+1} = B_i + \bar{B}_i.f(\bar{B}_i).T_i - Y_i$$

6.4 LONG–TERM PROJECTIONS (LT) (EQUILIBRIUM CONDITIONS)

The situation of equilibrium at the interval T_i implies that the biomass of the stock, at the end of the interval T_i (usually 1 year) is equal to the biomass at the beginning of the same interval, $B_{i+1}=B_i$, or the variation of the biomass is zero $\Delta B_i=0$.

Bringing the instantaneous rates closer to the mean rates, when the stock is in equilibrium, during T_i, then $\Delta B_i=0$ and $rmr(B_t)=0$. Thus, a equilibrium condition will be:

$$f.(\bar{B}_i) = F_i$$

Then the equilibrium conditions, referred to with the subindex E are:

$$f(\bar{B}_E) = F_E$$

$$Y_E = F_E.\bar{B}_E.T$$

6.5 BIOMASS AND FISHING LEVEL INDICES

In practice the values of $\bar{B}_i$ e F_i are not always available and so, one has to look for quantities that are associated with the biomass, B, and the fishing level, F, preferably quantities (called *indices*) proportional to those parameters.

Let $\bar{U}$ be an index of the mean biomass, $\bar{B}$, then during the interval of time T we have:

$$\bar{U} = q.\bar{B}$$

and let f be the index of fishing mortality coefficient, F_i, then during the interval of time T:

$$f = const.F.T$$

from $\quad Y = F.\bar{B}.T \quad$ and $\quad \bar{U} = q.\bar{B}$

will have $\quad Y = F.(1/q).\bar{U}.T$

thus, to have $Y = f.\overline{U}$ is necessary to be $$f = \frac{1}{q}.FT$$

A very common index of $\overline{B}$ is the catch per unit effort (cpue). The index of F will be the fishing effort in an appropriate unit, in order to be proportional to the fishing level.

The constant of proportionality, q , is designated as the capturability or catchability coefficient and indicates the fraction of the biomass that is caught by unit of effort.

6.6 BIOLOGICAL TARGET REFERENCE POINTS (TRP)

Long–term (or equilibrium) biological reference points can also be defined for these models.

F_{MSY} is the value of F that makes the long–term capture, Y, maximum.

F_{MSY} is different of F_{max}. In fact F_{MSY} maximizes the Catch in weight, while F_{max}, maximizes the Catch in weight per Recruit. Notice that the value of F_{max} cannot be calculated with production models, because the age structure of the stock and the recruitment, R are considered implicit in the basic assumptions of the model.

The biological reference points depend on the basic assumptions of the model , therefore the value of F_{MSY} of the structural models is different from the value of F_{MSY} of the production models because the relation S-R, as well as the natural mortality coefficient, M, are implicit in the production models.

To compare results of the two types of models one has to take into consideration that each model is based on different basic assumptions.

For the same reasons, $F_{0.1}$ of the production models is a different concept to $F_{0.1}$ of the structural models.

$F_{0.1}$, $B_{0.1}$ and $Y_{0.1}$ of the productions models could be calculated directly from the basic assumptions but it is preferable to obtain those characteristics using the constant relations between the reference points 0.1 and MSY (Cadima, 1991).

6.7 TYPES OF PRODUCTION MODELS

The most common production models in fishery stocks assessment are the Schaefer model (1954), the Fox model (1970) and the Pella and Tomlinson model (1969), the latter is also designated as GENPROD (name of the computer program that the authors elaborated for the application of their model). Fox mentions that the elaboration of his model was based on an idea from Garrod (1969).

Each one of these models corresponds to one particular function of $H(B_t)$ of the basic assumption.

6.7.1 SCHAEFER MODEL

The function $H(B_t)$ of the basic assumption of this model is:

$$H(B_t) = B_t^{-1}$$

Relative instantaneous rate,* rir(B_t)*, due to natural causes

The general basic assumption of the Schaefer model is:

$$rir_{natural}[k^{-1} - B^{-1}] = -r$$

and then, the instantaneous rate of variation of the "natural" biomass can be mathematically deduced as:

$$rir(B_t)_{natural} = r.\left[1 - \frac{B_t}{k}\right]$$

Equilibrium conditions

The relative mean rate , rmr(B_t), in relation to $\overline{B}$, will be:

$$\underset{\text{relative to }\overline{B_E}}{tmr(B_t)} = f(\overline{B_E}) = r.\left[1 - \frac{\overline{B_E}}{k}\right]$$

and, as in equilibrium, $f(\overline{B_E}) = F_E$, the equilibrium conditions can then be expressed as:

$$\overline{B_E} = k.(1 - F_E / r)$$
$$Y_E = F_E.\overline{B}_E.T_i$$

Notice that $\overline{B}_E$ is linear with F_E and for $F_E = 0$, $\overline{B}_E = k$ = carrying capacity = virgin biomass.

Graphically, the relation between $\overline{B}_E$ and F_E shows a straight line with interception equal to k and slope equal to -k/r.

***Target point,* F_{MSY}**

The Schaefer equilibrium conditions during one year are:

$$Y_E = F_E \,.\, \overline{B}_E \qquad\qquad \overline{B}_E = k \,.\, (1 - F_E / r)$$

Y maximum will occur when dY/dF=0, then derivating the previous expression of Y_E in order to F and making it equal to zero the target point F_{MSY} will be:

Target point, F_{MSY} (Schaefer)

$$F_{MSY} = r/2 \qquad B_{MSY} = k/2 \qquad Y_{MSY} = rk/4$$

In fact, the derivative

$$dY/dF = \bar{B} + F\,(dB/dF)$$

or

$$dY/dF = k(1 - F/r) + F\,(-k/r) = k - 2k.F/r$$

and then,

$$F_{MSY} = r/2$$

the relations of the remaining characteristics are obtained by substituting this result in the equilibrium conditions.

Target point, $F_{0.1}$

The ratio between $F_{0.1}$ and F_{MSY} is constant and equal to 0.90, so:

Target point, $F_{0.1}$ **(Schaefer)**

$F_{0.1}/F_{MSY} = 0.90$ $\qquad B_{0.1}/B_{MSY} = 1.10$ $\qquad Y_{0.1}/Y_{MSY} = 0.99$

In fact, as seen before, $dY/dF = k - 2k.F/r$ and, as $F_{0.1}$ corresponds to $dY/dF = 0.1k$, so:

$$0.1k = k - 2kF_{0.1}/r$$

or

$$0.90 = 2\,F_{0.1}/r$$

or

$$0.90 = F_{0.1}/F_{MSY}$$

Abundance indices, $\bar{U}$*, and fishing level indices,* f

As seen in Section 6.5, the indices $\bar{U}$ and f, are assumed to be proportional to $\bar{B}$ and F, so the equilibrium condition can be written as:

$\bar{U}_E = a + b.f_E$ and $Y_E = f_E\,.\,\bar{U}_E$ $\qquad$ (a,b are constants).

The target point, f_{MSY}, is obtained by equating to zero the derivative of Y_E in order to f_E:

Target point, f_{MSY} **(Schaefer)**

$f_{MSY} = -a/(2b)$ $\qquad \bar{U}_{MSY} = a/2$ $\qquad Y_{MSY} = -a^2/(4b)$

In the production models, the ratios $f_{0.1}/f_{MSY}$ e $\bar{U}_{0.1}/\bar{U}_{MSY}$ are equal to the ratios $F_{0.1}/F_{MSY}$ and $B_{0.1}/\bar{B}_{MSY}$. With Schaefer's model we will then have:

Target point, $f_{0.1}$ **(Schaefer)**

$f_{0.1}/f_{MSY} = 0.90$ $\qquad \bar{U}_{0.1}/\bar{U}_{MSY} = 1.10$ $\qquad Y_{0.1}/Y_{MSY} = 0.99$

From $\overline{U} = q.\ \overline{B}$, and FT=q.f , the previous expressions of F_{MSY} and f_{MSY} , one can also obtain the relations between the parameters k and r and the coefficients a, b and q:

$$k = a/q \qquad r = -\ aq\ /(bT) \qquad kr = -\ a^2/(bT)$$

When the value of the interval T is 1 year, T will not appear in these expressions. It is possible to calculate the parameters k and r, knowing the values of the capturability coefficient, q. Notice that the product k.r does not depend on q.

6.7.2 FOX MODEL

For the Fox production model the function $H(B_t)$ will be:

$$H(B_t) = \ln(B_t)$$

Relative instantaneous rate,* $rir(B_t)$*, due to natural causes

For the Fox model, from the expression of the general basic assumption, we have:

$$rir_{natural}\ [\ \ln k - \ln B\ ] = -\ r$$

and then, as previously referred to, the instantaneous rate of variation of the "natural" biomass can be mathematically deduced from that expression and written as:

$$rir(B_t)_{natural} = r.\ln(k / B_t)$$

Equilibrium conditions

The equilibrium condition of the biomass can be expressed by:

$$r.\ln(k / \overline{B}_E) = F_E$$

Then, the equilibrium conditions will be:

$$\overline{B}_E = k.e^{-F_E / r} \qquad Y_E = F_E.\overline{B}_E.T_i$$

Notice that $\ln(\overline{B}_E)$ is linear with F_E and that, for $F_E = 0$, $\overline{B}_E = k$ = virgin biomass or carrying capacity. The relation between $\ln(\overline{B}_E)$ and F_E is linear, with interception equal to lnk, and slope = -1/r.

***Target point,* F_{MSY}**

Derivating Y_E in order to F and equating the derivative to zero, F_{MSY}, B_{MSY} and Y_{MSY} will be:

$$F_{MSY} = r \qquad B_{MSY} = k/e \qquad Y_{MSY} = rk/e$$

Target point, $F_{0.1}$

In this model the ratio between $F_{0.1}$ and F_{MSY} is constant and equal to 0.7815. So, it can be written:

$$F_{0.1}/F_{MSY} = 0.7815 \qquad B_{0.1}/B_{MSY} = 1.2442 \qquad Y_{0.1}/Y_{MSY} = 0.9724$$

These results are obtained in a similar way to those for the Schaefer model. The equation to solve will be:

$$e^{-F_{0.1}/r}.(1 - F_{0.1}/r) = 0.1$$

which requires iterative methods to find the value of $F_{0.1}/r$. The solution is $F_{0.1}/r = 0.7815$ that is igual to $F_{0.1}/F_{MSY}$.

Abundance indices, $\bar{U}$ *, and fishing level indices,* f

For the Fox model the equilibrium condition can be written as:

$$\bar{U}_E = e^{a+b.f_E} \qquad \text{or} \qquad \ln \bar{U}_E = a + b.f \qquad \text{(a,b are constants)}$$

and

$$Y_E = f_E \,.\, \bar{U}_E$$

Target point, f_{MSY}

The target point , f_{MSY}, can be obtained by equating to zero the derivative of Y_E in order to f_E:

$$f_{MSY} = -1/b \qquad \bar{U}_{MSY} = e^a/e \qquad Y_{MSY} = -\, e^a/be$$

Target point, $f_{0.1}$

In the Fox model, the ratios $f_{0.1}/f_{MSY}$ et $\bar{U}_{0.1}/\ \bar{U}_{MSY}$ are equal to $F_{0.1}/F_{MSY}$ and $\bar{B}_{0.1}/\ \bar{B}_{MSY}$ and then:

$$f_{0.1}/f_{MSY} = 0.7815 \qquad \bar{U}_{0.1}/\ \bar{U}_{MSY} = 1.2442 \qquad Y_{0.1}/Y_{MSY} = 0.9724$$

From $\bar{U} = q.\bar{B}$ and from FT=q.f , the following can be deduced:

$$k = e^a/q. \qquad r = -\, q/(bT) \qquad kr = -\, e^a/(bT)$$

When the value of the interval T is one year, T will not appear in those expressions. The last expression allows the calculation of the product k.r. To calculate k and r separately it is necessary to know the value of the coefficient of capturability, q.

6.7.3 PELLA AND TOMLINSON MODEL (GENPROD)

For this production model the function $H(B_t)$ will be:

$$H(B_t) = B_t^{-p}$$

Relative instantaneous rate,* rir(B_t)*, due to natural causes

The expression of the basic assumption of the GENPROD model, will be :

$$rir_{natural}\left(k^{-p} - B^{-p}\right) = -r$$

therefore

$$rir(B_t)_{natural} = (r/p).[1-(k/\overline{B})^{-p}]$$

Equilibrium conditions

In equilibrium conditions, F_E will be:

$$F_E = f(\overline{B}_E) = (r/p).[1-(k/\overline{B})^{-p}]$$

Then, the equilibrium conditions can be expressed as:

$$\overline{B}_E = k.[1-p.F_E/r]^{1/p} \quad Y_E = F_E.\overline{B}_E.T_i$$

Notice that the relation between $(\overline{B}_E)^p$ and F_E is linear with intercept equal to k and the slope equal to -pk/r, in conclusion, for $F_E = 0$, $\overline{B}_E = k$ = carrying capacity=virgin biomass.

***Target point,* F_{MSY}**

Derivating $\mathbf{Y_E}$ in order to **F** and equating to zero, we will have:

Target point,* F_{MSY} *(Pella and Tomlinson)

$$F_{MSY} = r.\left[\frac{1}{1+p}\right] \qquad B_{MSY} = k.\left[\frac{1}{1+p}\right]^{\frac{1}{p}} \qquad Y_{MSY} = r.k.\left[\frac{1}{1+p}\right]^{\left(\frac{1}{p}+1\right)}$$

***Target point,* $F_{0.1}$**

The ratio between $F_{0.1}$ and F_{MSY} is constant for each value of p and can be obtained in a similar way to the previous cases. The equation to solve by iterative methods is:

$$X = 1 - 0.1.[1-p/(1+p).X]^{1/p} \qquad \text{where} \qquad X = F_{0.1}/F_{MSY}$$

And also $\qquad B_{0.1}/B_{MSY} = [1+p - p./(1+p).X]^{(1+1/p)}$

$$Y_{0.1}/Y_{MSY} = [F_{0.1}/F_{MSY}].[B_{0.1}/B_{MSY}]$$

The following Table summarizes the most important results:

	parameter p	$F_{0.1}/F_{MSY}$	$B_{0.1}/B_{MSY}$	$Y_{0.1}/Y_{MSY}$
Fox	0.0 *	0.781521	1.244182	0.972355
	0.2	0.819995	1.193441	0.978616
	0.4	0.848355	1.158613	0.982915
	0.6	0.869888	1.133469	0.985991
	0.8	0.886657	1.114599	0.988268
Schaefer	1.0 *	0.900000	1.100000	0.990000
	1.2	0.910816	1.088420	0.991350
	1.4	0.919724	1.079045	0.992424
	1.6	0.927165	1.071323	0.993293
	1.8	0.933457	1.064867	0.994008
	2.0	0.938835	1.059401	0.994602
	2.2	0.94377	1.054720	0.995704
	2.4	0.947516	1.050674	0.995531
	2.6	0.951059	1.047146	0.995898
	2.8	0.954188	1.044045	0.996216
	3.0	0.956969	1.041302	0.996494

Notice that $(F_{0.1}/F_{MSY}) + (B_{0.1}/B_{MSY}) \cong 2$. From this result it can be said that when $F_{0.1}$ is smaller than F_{MSY} by a certain percentage, the equivalent relation of the biomasses will be bigger by the same percentage.

Abundance indices, $\overline{U}$ *, and fishing level indices,* f

For the Pella and Tomlinson model, the equilibrium conditions can be written as:

$$\overline{U} = (a + b.f)^{1/p} \quad \text{or} \quad \overline{U}^{p} = (a + b.f) \text{ (a,b are constants).}$$

$$Y_E = f_E \, . \, \overline{U}_E$$

The target point, f_{MSY}, can be obtained by equating to zero the derivative of Y_E in order to f_E:

Target point, f_{MSY}

$$f_{MSY} = -a/(b(1+1/p)) \qquad \overline{U}_{MSY} = [a/(1+p)]^{1/p} \qquad Y_{MSY} = -(p/b).(a/(1+p))^{(1+1/p)}$$

The ratios $f_{0.1}/f_{MSY}$ and $\overline{U}_{0.1}/\overline{U}_{MSY}$ will be equal to the ratios $F_{0.1}/F_{MSY}$ and $\overline{B}_{0.1}/\overline{B}_{MSY}$, respectively. These last ratios can be observed in the previous Table.

The values of k, r e kr can also be obtained from $\overline{U} = q.\ \overline{B}$, and from F.T=q.f

$k = a^{1/p}/q$ $r/= -\ apq\ /(bT)$ $kr = (-p/bT)a^{(1+1/p)}$

When the size of the interval is one year, T will not appear in those expressions. The last expression in the previous table allows the calculation of the product k.r.The separate values of k and r can be calculated if the value of the coefficient of capturability, q is known.

Comments

1. The Pella and Tomlinson model has been criticized in its practical application because sometimes it produces better adjustments with non reliable values of the parameter p, resulting in extremely high values of F_{MSY}.
2. It is also important to notice that p, the additional parameter of that model, is written with different symbols depending on the authors.
3. The values of the biological reference points relative to F, estimated by Schaefer's model, are more restrictive than the corresponding values estimated by the Fox or GENPROD production models.

6.8 SHORT–TERM PROJECTIONS

6.8.1 GENERAL METHODS

Long–term projections have been estimated in fisheries since the 50's using these production models but in practice, it was only in the 90's that methods were developed for short–term projections. These methods are based in the Schaefer, Fox and Pella et Tomlinson expressions for the non-exploited biomass.

By applying production models as referred to in Section 6.3, the variation of the biomass for 1 year can be expressed, in a general way, as:

$$\Delta B_{i\,total} \cong f(\overline{B}_i).\overline{B}_i.T_i - Y_{i_{fishing}}$$

or $$B_{i+1}=B_i + \overline{B}_i.f(\ \overline{B}_i\ ).T_i - Y_i$$

where B_i = biomass at the beginning of the year i

B_{i+1}= biomass at the end of the year i

$\overline{B}_i$ = mean biomass during the year i

Y_i = catch in weight during the year i

$f(\overline{B}_i)$ is the approximation of the mean rate of "natural" variation of the biomass, relative to $\overline{B}_i$ during the year i.

The expression of the variation of the biomass is the basis for most of the methods for short–term projections. Computer programs were prepared for the application of these methods, which also determine long–term projections, biological reference points, etc. Some examples are CEDA and BYODIN, Rosenberg *et al.* (1990), and Punt and Hilborn (1996) respectively.

Theoretically those methods suppose that $\bar{B}_i$ et Y_i are known for a period of years. The function f($\bar{B}_i$) can be that of Schaefer, Fox or Pella and Tomlinson.

To determine the parameters r and k it would be necessary to adopt one of the expressions of f($\bar{B}_i$) and the value B_1 in the first interval of the period of years.

In practice, the values of the annual mean biomasses are not available, only the associated quantities, usually assumed to be proportional to the mean biomasses, that is indices $U_i=q.\ \bar{B}_i$ so, the parameters to be estimated are r, k and q (see Chapter 7).

The object function, of the least squares method, to be minimized is $\Phi = \Sigma\ (U_{obs} - U_{mod})^2$ that is, the sum of the squares of the residuals between the observed values and the estimated values (*designated by error of the process*). However, when the the relation $U=q.\ \bar{B}_i$ is not determinant, but it is supposed to have an error, *designated as observation error*, then it is preferable to adopt the object function $\Phi = \Sigma\ (\ln U_{obs} - \ln U_{mod})^2$ (Punt and Hilborn, 1996).

6.8.2 PRAGER METHOD (1994)

Prager (1994) adopted the Schaefer model and used the relative *instantaneous* rate of the variation of the biomass in the initial basic expression (not the *mean* rate approximation) that is,

$$rir(B_t) = r[1 - B_t / k].\ B_t - F_i.\ B_i$$

He integrated this expression during the year i and obtained the relation between B_{i+1} and B_i

He also calculated the mean biomass, $\bar{B}_i$, integrating B_t during the year i. Finally the catch in weight is calculated as:

$$Y_i = F_i\ \bar{B}_i\ .$$

The estimation of the parameters can then be made using the least squares method. The computer program prepared for this estiamtion is called ASPIC (Prager, 1995).

6.8.3 YOSHIMOTO AND CLARKE METHOD (1993)

The short–term projections are derived from the basic assumption of the production models,

$$rir(B_t)_{total} = rir(B_t)_{natural} - (F_t)_{fishing}$$

or, representing $rir(B_t)_{natural}$ by $f(B_t)$:

$$rir(B_t)_{total} = f(B_t) - (F_t)_{fishing}$$

Integrating this expression during the interval of time T_i and considering that:

$$rir(B_t) = air\ \ln(B_t) \quad \text{and} \quad F_t = F_i = \text{constant:}$$

$$\int_{ti}^{ti+1} air(\ln B_t)dt =$$

$$\ln(B_{i+1}) - \ln(B_i) = f(B_t)_{cum_i} - F_i\,T_i = \overline{f}(B_t)\,T_i - F_i\,T_i$$

For the next interval T_{i+1} (which is the interval where one intends to project the stock and the catch):

$$\ln(B_{i+2}) - \ln(B_{i+1}) = \overline{f}(B_{t+1}).T_{i+1} - F_{i+1}.T_{i+1}$$

Calculating the simple arithmetic mean of the two previous expressions and considering that :

$$1/2[\ln(B_{i+1}) + \ln(B_{i+2})] - 1/2[\ln(B_i) + \ln(B_{i+1})] = \ln(\overline{B}^*_{i+1}) - \ln(\overline{B}^*_i)$$

where $\overline{B}^*_i$ is the geometric mean of B_i and B_{i+1}, and $\overline{B}^*_{i+1}$ is the geometric mean of B_{i+1} and B_{i+2},

Therefore, the mean of the two expressions will be:

$$\ln(\overline{B}^*_{i+1}) - \ln(\overline{B}^*_i) = (1/2)\{\overline{f}(B_i)\,T_i + \overline{f}(B_{i+1})\,T_{i+1}\} - (1/2)\{F_i\,T_i + F_{i+1}\,T_{i+1}\}$$

The natural rir(Bt) of the Fox model is, as mentioned before $f(B_t) = r(\ln k - \ln B_t)$, so the approximation, $\overline{f}(B_i)$ can be written as:

$$r[\ln k - \ln(\overline{B}^*_i)]$$

where $\overline{B}^*_i$ is the geometric mean of B_i and B_{i+1}.

Therefore, the previous expression relative to the geometric means, can be re-written as:

$$\ln(\overline{B}^*_{i+1}) - \ln(\overline{B}^*_i) =$$

$$= (1/2)\{r(\ln k - \ln(\overline{B}^*_i))\,T_i + r(\ln k - \ln(\overline{B}^*_{i+1}))\,T_{i+1}\} - (1/2)\{F_i\,T_i + F_{i+1}\,T_{i+1}\}$$

To simplify, and as the intervals of T_i are usually constant (and equal to one year), one can use T instead of T_i and T_{i+1} and the expression will be:

$$\ln(\overline{B}^*_{i+1}) - \ln(\overline{B}^*_i) = (rT/2)\{\ln k - \ln(\overline{B}^*_i) + \ln k - \ln(\overline{B}^*_{i+1})\} - (1/2)\{F_i\,T + F_{i+1}\,T\}$$

or reorganizing the terms of this expression, it will be:

$$(1 + rT/2)\,.\,\ln(\overline{B}^*_{i+1}) = rT\,\ln k + (1 - rT/2)\,\ln(\overline{B}^*_i)) - (T/2)(F_i + F_{i+1})$$

Finally, the expression can be written as follows:

$$\ln(\bar{B}^*_{i+1}) = 2rT/(2+rT) \,.\, \ln k + (2-rT)/(2+rT) \,.\, \ln(\bar{B}^*_i) - T/(2+rT) \,.\, (F_i + F_{i+1})$$

As seen in the long–term projections (or equilibrium), it is more common to have biomass indices, $\bar{U}$, and fishing level indices, f, rather than $\bar{B}$ et F values.

Using the indices $\bar{U} = q\,\bar{B}$ and $qf = FT$

the Yoshimoto and Clarke expression (1993) can be written as :

$$\ln \bar{U}_{i+1} = \frac{2rT}{2+rT}.\ln(qk) + \frac{2-rT}{2+rT}.\ln \bar{U}_i - \frac{q}{2+rT}(f_i + f_{i+1})$$

It is useful, in practice, to write this expression in the following way:

$$\ln\,\bar{U}_{i+1} = b_1 + b_2 \,.\, \ln\,\bar{U}_i + b_3 \,.\, (f_i + f_{i+1})$$

where:

$$b_1 = \frac{2rT}{2+rT}\ln(qk)$$

$$b_2 = \frac{2-rT}{2+rT}$$

$$b_3 = -\frac{q}{2+rT}$$

From the coefficients b_1, b_2 and b_3 one can estimate the parameters q, r and k (keep in mind that in the long–term projections it was not possible to obtain q separately) as :

$$q = -4b3/(1+b2)$$

$$rT = 2(1-b_2)\,/\,(1+b_2)$$

$$k = ((1+b_2)/(-4b_3)).e^{b_1/(1-b_2)}$$

Comments

1. The fact of having developed, in this manual, the Yoshimoto and Clarke model for the short–term projections, does not mean a special preference for this model over other models for the short–term projections.
2. Yoshimoto and Clarke designated their expression by the integrated expression of Fox, as it is based on the direct integration of the basic assumption.

3. Notice that $\overline{f}(B_i)$ and $\overline{f}(B_{i+1})$ are, in general, different from $f(\overline{B}_i)$ and $f(\overline{B}_{i+1})$. However, the means of f( ) may be considered equal to f(means of B) if another type of mean of B is used.

- Definition of $\overline{B}_i^*$ through a function

  Consider n values B_i and the simple arithmetic mean of $f(B_i)$, that is,

$$\overline{f(B)} = (1/n).\Sigma f(B_i)$$

- Let $\overline{B}^*$ be a value such as $f(\overline{B}^*) = \overline{f}(B)$.

  $\overline{B}^*$ is designated as the mean of B_i through the function f.

EXAMPLES

4. If $f(B) = \ln(B)$ then $\ln(\overline{B}^*) = (1/n).\Sigma \ln(B_i)$ and $\overline{B}^*$ is the mean of the values B_i through the logarithm function, also designated as geometric mean of the values B_i
5. If $f(B) = B^{-1}$ then $(\overline{B}^*)^{-1} = (1/n).\Sigma(B_i)^{-1}$ and $\overline{B}^*$ is designated as harmonic mean of the values B_i
6. If $f(B) = B^{-p}$ then $(\overline{B}^*)^{-p} = (1/n).\Sigma(B_i)^{-p}$ and $\overline{B}^*$ is designated as the mean of order (-p) of the values B_i .
7. Another approach (Cadima & Pinho, 1995) of the integrated equation of Fox can be:

$$\ln \overline{U}_{i+1} = b_1 + b_2 . \ln \overline{U}_i + b_3 . (f_i + f_{i+1})$$

where:

$$b_1 = (1-e^{-rT}) \ln(qk) \qquad k = \frac{1-b_2}{-2 \cdot b_3 \cdot \ln b_2} \cdot e^{\frac{b_1}{1-b_2}}$$

$$b_2 = e^{-rT} \qquad rT = -\ln b_2$$

$$b_3 = -q(1-e^{-rT}) / (2rT) \qquad q = 2 \cdot \ln b_2 \cdot \frac{b_3}{1-b_2}$$

This last approach of the integrated Fox model can be deduced from the basic assumption of the model, during the interval T_i:

$$rir(B_t)_{total} = r.(\ln k - \ln B_t)_{naturalFox} - (F_i)_{fishing}$$

Taking into account the properties of rates and assuming r, k and F_i constant during T_i interval, the absolute instantaneous rate of $[r.(\ln K - \ln B_t) - F_i]$ will be:

$$air[r.(\ln k - \ln B_t) - F_i] = -r.air(\ln B_t) = -r.rir(B_t)$$

So substituting rir (B_t) by the Fox expression mentioned before, one can write:

$$air[r.(\ln k - \ln B_t) - F_i] = -r. [r.(\ln k - \ln B_t) - F_i]$$

or $$\frac{air[\,r.(\ln k - \ln B_t) - F_i\,]}{[\,r.(\ln k - \ln B_t) - F_i\,]} = -r$$

Finally, by the definition of rir the expression will be:

$$rir[\,r.(\ln k - \ln B_t) - F_i\,] = -r$$

showing that $[r.(\ln k - \ln B_t)-F_i]$ follows an exponential model, during the interval T_i with r constant.

So, the final value of $(r.\ln k - r.\ln B_t - F_i)$, can be expressed as:

$$(r.\ln k - r.\ln B_{i+1} - F_i) = (r.\ln k - r.\ln B_i - F_i).\ e^{-r.Ti}$$

or

$$\ln B_{i+1} = (1- e^{-r.Ti}).\ln k + e^{-r.Ti}.\ln B_i - (1- e^{-r.Ti}).F_i/r$$

At the following interval, T_{i+1}, the expression would be:

$$\ln B_{i+2} = (1- e^{-r.Ti+1}).\ln k + e^{-r.Ti+1}.\ln B_{i+1} - (1- e^{-r.Ti+1}).F_{i+1}/r$$

then, the mean of the two previous expressions, considering $T_i=T_{i+1}=T$, will be:

$$\ln\ \overline{B}_{i+1}^{\,*} = (1- e^{-r.T}).\ln k + e^{-r.T}.\ \ln\ \overline{B}_i^{\,*} - ((1- e^{-r.T}) / 2r).(F_i+F_{i+1})$$

where $\overline{B}_i^{\,*}$ = geometric mean of B_i and B_{i+1} and $\overline{B}_{i+1}^{\,*}$ = geometric mean of B_{i+1} and B_{i+2}.

Using the indices $\overline{U}_i = q.\ \overline{B}_i^{\,*}$ and $qf_i = F_iT$ the expression will be:

$$\ln\ \overline{U}_{i+1} = (1- e^{-r.T}).\ln (qk) + e^{-r.T}.\ \ln\ \overline{U}_i - ((1- e^{-r.T})/2rT)\ .q(f_i+f_{i+1})$$

which is the initial expression of comment nº 7.

CHAPTER 7 – ESTIMATION OF PARAMETERS

In the previous chapters, several models used in stock assessment were analysed, the respective parameters having been defined. In the corresponding exercises, it was not necessary to estimate the values of the parameters because they were given. In this chapter, several methods of estimating parameters will be analysed. In order to estimate the parameters, it is necessary to know the sampling theory and statistical inference.

This manual will use one of the general methods most commonly used in the estimation of parameters – the least squares method. In many cases this method uses iterative processes, which require the adoption of initial values. Therefore, particular methods will also be presented, which obtain estimates close to the real values of the parameters. In many situations, these initial estimates also have a practical interest. These methods will be illustrated with the estimation of the growth parameters and the S-R stock-recruitment relation.

The least squares method is presented under the forms of Simple linear Regression, multiple linear model and non linear models (method of Gauss-Newton).

Subjects like residual analysis, sampling distribution of the estimators (asymptotic or empiric Bookstrap and jacknife), confidence limits and intervals, etc., are important. However, these matters would need a more extensive course.

7.1 SIMPLE LINEAR REGRESSION – LEAST SQUARES METHOD

Model

Consider the following variables and parameters:

Response or dependent variable	= Y
Auxiliary or independent variable	= X
Parameters	= A,B

The *response variable is linear with the parameters*

$$Y = A+BX$$

Objective

The objective of the method is to estimate the parameters of the model, based on the observed pairs of values and applying a certain criterium function (the observed pairs of values are constituted by selected values of the auxiliary variable and by the corresponding observed values of the response variable), that is :

Observed values	x_i and y_i for each pair i, where i=1,2,...,i,...n
Values to be estimated	A and B and $(Y_1, Y_2, ..., Y_i, ..., Y_n)$ for the n observed pairs of values

(Estimates values: $\hat{A}$ and $\hat{B}$ (or a and b) and $(\hat{Y}_1, \hat{Y}_2, ..., \hat{Y}_i, ..., \hat{Y}_n)$)

Object function (or criterium function)

$$\Phi = \sum_{i=1}^{n} (y_i - Y_i)^2$$

Estimation method

In the least squares method the estimators are the values of A and B which minimize the object function. Thus, one has to calculate the derivatives $\partial\Phi/\partial A$ e $\partial\Phi/\partial B$, equate them to zero and solve the system of equations in A and B.

The solution of the system can be presented as :

$$\bar{x} = (1/n).\Sigma x \qquad \bar{y} = (1/n).\Sigma y$$

$$Sxx = \Sigma(x - \bar{x})(x - \bar{x}) \qquad Sxy = \Sigma(x - \bar{x})(y - \bar{y})$$

$$b = Sxy/Sxx \qquad a = \bar{y} - b.\ \bar{x}$$

Notice that the observed values y, for the same set of selected values of X, depend on the collected sample. For this reason, the problem of the simple linear regression is usually presented in the form :

$$y = A + BX + \varepsilon$$

where ε is a random variable with *expected value* equal to zero and *variance* equal to σ^2. So, the expected value of y will be Y or A+BX and the variance of y will be equal to the variance of ε.

The terms deviation and residual will be used in the following ways:

Deviation is the difference between $y_{observed}$ and y_{mean} ($\bar{y}$) i.e., deviation = $(y - \bar{y})$

while

Residual is the difference between $y_{observed}$ and $Y_{estimated}$ ($\hat{Y}$), i.e., residual = $(y_i - \hat{Y}_i)$.

To analyse the adjustment of the model to the observed data, it is necessary to consider the following characteristics:

Sum of squares of the residuals:

$$SQ_{residual} = \Sigma\left(y - \hat{Y}\right)^2$$

This quantity indicates the residual variation of the observed values in relation to the estimated values of the response variable of the model, which can be considered as the variation of the observed values that is not explained by the model.

Sum of squares of the deviations of the estimated values of the response variable of the model:

$$SQ_{model} = \Sigma\left(\hat{Y} - \bar{y}\right)^2$$

This quantity indicates the variation of the estimated values of the response variable of the model in relation to its mean, that is the *variation* of the response variable explained by the model.

Total sum of squares of the deviations of the observed values equal to:

$$SQ_{residual} = \sum\left(y - \bar{y}\right)^2$$

This quantity indicates the total variation of the observed values in relation to the mean

It is easy to verify the following relation:

$$SQ_{total} = SQ_{model} + SQ_{residual}$$

or

$$1 = \frac{SQ_{model}}{SQ_{total}} + \frac{SQ_{residual}}{SQ_{total}}$$

or $1 = r^2 + (1 - r^2)$

where

r^2 (coefficient of determination) is the percentage of the total variation that is explained by the model and

$1 - r^2$ is the percentage of the total variation that *is not explained* by the model.

7.2 MULTIPLE LINEAR REGRESSION – LEAST SQUARES METHOD

Model

Consider the following variables and parameters:

Response or dependent variable	$= Y$
Auxiliary or independent variables	$= X_1, X_2,..., X_j,..., X_k$
Parameters	$= B_1, B_2,..., B_j,..., B_k$

The response variable is linear with the parameters

$$Y = B_1X_1+B_2X_2+... + B_kX_k = \Sigma B_jX_j$$

Objective

The objective of the method is to estimate the parameters of the model, based on the observed **n** sets of values and by applying a certain criterium function (the observed sets of values are constituted by selected values of the auxiliary variable and by the corresponding observed values of the response variable), that is:

Observed values $x_{1,i}\ x_{2,i}\ ,..., x_{j,i,..}, x_{k,i}$ and y_i for each set i, where i=1,2,...,i,...n

Values to be estimated $B_1,B_2,...,B_j,...,B_k$ et $(Y_1,Y_2,..., Y_i ,..., Y_n)$

The estimated values can be represented by :

$$\hat{B}_1,\ \hat{B}_2 ,..., \hat{B}_j ,..., \hat{B}_k \text{ (ou } b_1,b_2,...,b_j,...,b_k\text{) et } \hat{Y}_1,\ \hat{Y}_2 ,..., \hat{Y}_i ,..., \hat{Y}_n$$

Object function (or criterium function)

$$\Phi = \sum_{i=1}^{n} (y_i - Y_i)^2$$

Estimation method

In the *least squares method* the estimators are the values of B_j which minimize the object function.

As with the simple linear model, the procedure of minimization requires equating the partial derivatives of Φ to zero in order to each parameter, B_j, where j=1, 2, ..., k. The system is preferably solved using matrix calculus.

Matrix version

Matrix $X_{(n,k)}$ = Matrix of the n observed values of each of the k auxiliary variables

Vector $y_{(n,l)}$ = Vector of the n observed values of the response variable

Vector $Y_{(n,l)}$ = Vector of the values of the response variable given by the model (unknown)

Vector $B_{(k,l)}$ = Vector of the parameters

Vector $\hat{B}$ or $b_{(k,l)}$ = Vector of the estimators of the parameters

Model

$$Y_{(n,1)} = X_{(n,k)} \cdot B_{(k,1)} \text{ ou } Y=X.B+\varepsilon$$

Object function

$$\Phi_{(1,1)} = (y-Y)^T.(y-Y) \quad \text{ou} \quad \Phi_{(1,1)} = (y-X.B)^T.(y-X.B)$$

To calculate the least squares estimators it will suffice to put the derivative $d\Phi/dB$ of Φ in order to vector B, equal to zero. $d\Phi/dB$ is a vector with components $\partial\Phi/\partial B_1$, $\partial\Phi/\partial B_2$, ..., $\partial\Phi/\partial B_k$. Thus:

$$d\Phi/dB_{(k,1)} = -2.X^T.(y-X.B) = 0$$

or $$X^Ty - (X^T.X).\ B = 0$$

and $$b = \hat{B} = (X^T.X)^{-1}.\ X^Ty$$

The results can be written as:

$$b_{(k,1)} = (X^T.X)^{-1}.X^Ty$$

$$\hat{Y}_{(n,l)} = X.b \quad \text{or} \quad \hat{Y}_{(n-l)} = X\,(X^T.X)^{-1}.X^T y$$

$$\text{residuals}_{(n,1)} = (y-\hat{Y})$$

Comments

In statistical analysis it is convenient to write the estimators and the sums of the squares using idempotent matrices. Then the idempotent matrices L, (I - L) and (I - M) with

$L_{(n,n)} = X\,(X^T . X)^{-1} . X^T$, I = unity matrix and $M_{(n,n)}$ = $mean_{(n,1)}$ matrix = 1/n [1] where [1] is a matrix with all its elements equal to one, are used.

It is also important to consider the sampling distributions of the estimators assuming that the variables ε_i are independent and have a normal distribution.

A summary of the main properties of the expected value and variance of the estimators is presented :

$$E[c_1+c_2.u] = c_1+c_2.E[u] \qquad V[c_1+c_2.u] = c_2.V[u].c_2^T$$

1 –	Random variable, ε	$\varepsilon_{n.}$ (independent)
	Expected value of ε	$E[\varepsilon] = 0.$
	Variance of ε	$V[\varepsilon]_{(n.n)} = E[\varepsilon.\varepsilon^T] = I.\sigma^2$
2 –	Observed response variable y	$y = Y+\varepsilon$
	Expected value of y	$E[y] = Y = X.B.$
	Variance of y	$V[y]_{(n.n)} = V[\varepsilon]_{(n.n)} = I.\sigma^2$
3 –	Estimator of B	$\hat{B} = (X^T.X)^{-1}.X^T.y$
	Expected value of $\hat{B}$	$E[\hat{B}] = B$
	Variance of $\hat{B}$	$V[\hat{B}]_{(k.k)} = (X^T.X)^{-1}.\sigma^2$
4 –	Estimator of Y of the model	$\hat{Y} = X.\ \hat{B} = L.y$
	Expected value of $\hat{Y}$	$E[\hat{Y}] = Y.$
	Variance of $\hat{Y}$	$V[\hat{Y}] = L.\sigma^2$
5 –	Residual e	$e = y - \hat{Y} = (I-L).y$
	Expected value of e	$E[e] = 0$
	Variance of e	$V[e] = (I-L).\sigma^2$

6 – Sum of squares

6.1 - Residual Sum of squares = SQ $residual_{(1.1)} = (y-\hat{Y})^T(y-\hat{Y}) = y^T\,(I-L)y$

This quantity indicates the residual variation of the observed values in relation to the estimated values of the model, that is, the variation not explained by the model.

6.2 - Sum of squares of the deviation of the model = SQ $model_{(1.1)} = (\hat{Y}-\bar{y})^T(\hat{Y}-\bar{y}) = y^T\,(L-M)y$

This quantity indicates the variation of the estimated response values of the model in relation to the mean, that is, the variation explained by the model.

6.3 - Total Sum of the squares of the deviations = SQ total$_{(1.1)}$ = $(y-\bar{y})^T(y-\bar{y}) = y^T(I-M)y$

This quantity indicates the total variation of the observed values in relation to the mean.
It is easy to verify the following relation:

$$SQ_{total} = SQ_{model} + SQ_{residual} \quad \text{or}$$

$$1 = \frac{SQ_{model}}{SQ_{total}} + \frac{SQ_{residual}}{SQ_{total}}$$

or $$1 = R^2 + (1 - R^2)$$

where:

R^2 is the percentage of the total variation that is *explained* by the model. In matrix terms it will be:

$$R^2 = [y^T(L - M)y].[(y^T(I - M)y]^{-1}$$

$1-R^2$ is the percentage of the total variation that is not explained by the model.

The ranks of the matrices (I-L), (I-M) and (L-M) respectively equal to (n-k), (n-1) and (k-1), are the degrees of freedom associated with the respective sums of squares.

7.3 NON-LINEAR MODEL – METHOD OF GAUSS-NEWTON – LEAST SQUARES METHOD

Model

Consider the following variables and parameters:

Response or dependent variable = Y

Auxiliary or independent variable = X

Parameters = $B_1, B_2, ..., B_j, ..., B_k$

The response variable is non-linear with the parameters

Y = f(X;B) where B is a vector with the components $B_1, B_2, ..., B_j, ..., B_k$

Objective

The objective of the method is to estimate the parameters of the model, based on the n observed pairs of values and by applying a certain criterium function (the observed sets of values are constituted by selected values of the auxiliary variable and by the corresponding observed values of the response variable), that is:

Observed values x_i and y_i for each pair i, where i=1,2,...,i,...n

Values to be estimated $B_1,B_2,...,B_j,..,B_k$ and $(Y_1,Y_2,...,Y_i,...,Y_n)$ form the n pairs of observed values.

(Estimates = $\hat{B}_1, \hat{B}_2,..., \hat{B}_j,..., \hat{B}_k$ or $b_1,b_2,...,b_j,...,b_k$ and $\hat{Y}_1, \hat{Y}_2,..., \hat{Y}_i,..., \hat{Y}_n$)

Object function or criterium function

$$\Phi = \sum_{i=1}^{n}(y_i.Y_i)^2$$

Estimation criterium

The estimators will be the values of B_j for which the object function is minimum. (This criterium is called the least squares method).

Matrix version

It is convenient to present the problem using matrices.
So:

Vector $X_{(n,1)}$ = Vector of the observed values of the auxiliary variable

Vector $y_{(n,1)}$ = Vector of the observed values of the response variable

Vector $Y_{(n,1)}$ = Vector of the values of the response variable given by the model

Vector $B_{(k,1)}$ = Vector of the parameters

Vector $b_{(k,1)}$ = Vector of the estimators of the parameters

Model

$$Y_{(n,1)} = f(X; B)$$

Object function

$$\Phi_{(1,1)} = (y\text{-}Y)^T.(y\text{-}Y)$$

In the case of the non linear model, it is not easy to solve the system of equations resulting from equating the derivative of the function Φ in order to the vector B, to zero. Estimation by the least squares method can, based on the Taylor series expansion of function Y, use iterative methods.

Revision of the Taylor series expansion of a function

Here is an example of the expansion of a function in the Taylor series in the case of a function with one variable.

The approximation of Taylor means to expand a function $Y = f(x)$ around a selected point, x_0, in a power series of x :

$$Y = f(x) = f(x_0) + (x-x_0).f'(x_0)/1! + (x-x_0)^2 f''(x_0)/2! + ... + (x-x_0)^i f^{(i)}(x_0)/i! + ...$$

where $f^{(i)}(x_0)$ = i^{th} derivatives of f(x) in order to x, at the point x_0.

The expansion can be approximated to the desired power of x. When the expansion is approximated to the power 1 it is called a linear approximation, that is,

$$Y \cong f(x_0) + (x-x_0).f'(x_0)$$

The Taylor expansion can be applied to functions with more than one variable. For example, for a function $Y = f(x_1,x_2)$ of two variables, the linear expansion would be:

$$Y \approx f(x_{1(0)}, x_{2(0)}) + (x_1 - x_{1(0)}).\frac{\delta f(x_{1(0)}, x_{2(0)})}{\delta x_1} + (x_2 - x_{2(0)}).\frac{\delta f(x_{1(0)}, x_{2(0)})}{\delta x_2}$$

which may be written, in matrix notation, as

$$Y = Y_{(0)} + A_{(0)}.(x - x_{(0)})$$

where $Y_{(0)}$ is the value of the function at the point $x_{(0)}$,with components $x_{1(0)}$ and $x_{2(0)}$,and $A_{(0)}$ is the matrix of derivatives whose elements are equal to the partial derivatives of $f(x_1,x_2)$ in order to x_1,x_2 at the point $(x_{1(0)}, x_{2(0)})$.

To estimate the parameters, the Taylor series expansion of function Y is made in order to the parameters B and not to the vector X.

For example, the linear expansion of $Y = f(x,B)$ in $B_1, B_2, ..., B_k$, would be:

$$Y = f(x;B) = f(x; B_{(0)}) + (B_1 - B_{1(0)})\, \partial f/\partial B_1 (x;B_{(0)}) + ... +$$
$$(B_2 - B_{2(0)})\, \partial f/\partial B_2 (x;B_{(0)}) + ... + ... + (B_k - B_{k(0)})\, \partial f/\partial B_k (x;B_{(0)})$$

or, in matrix notation, it would be :

$$Y_{(n,1)} = Y_{(0)\,(n,1)} + A_{(0)\,(n,k)} \,.\, \Delta B_{(0)\,(k,1)}$$

where

A = matrix of order (n,k) of the partial derivatives of the matrix f(x;B) *in order to the vector* B *at the point* $B_{(0)}$ and

$$\Delta B_{(0)} = \text{vector } (B - B_{(0)}).$$

Then, the object function will be:

$$\Phi = (y-Y)^T.(y-Y) = (y - Y_{(0)} - A_{(0)}.\Delta B_{(0)})^T (y - Y_{(0)} - A_{(0)}.\Delta B_{(0)})$$

To obtain the minimum of this function it is more convenient to differentiate Φ in order to the vector ΔB *than in relation to vector* B *and put it equal to zero*. Thus:

$$0 = -2(A_{(0)})^T(y-Y_{(0)} - A_{(0)}.\Delta B_{(0)}) = -2A_{(0)}^T(y-Y_{(0)}) + 2\,A^T_{(0)}\,A_{(0)}.\Delta B_{(0)}$$

or

$$A^T_{(0)}A_{(0)}.\Delta B_{(0)} = A^T_{(0)}(y-Y_{(0)})$$

Therefore:

$$\Delta B_{(0)} = \left(A^T_{(0)}.A_{(0)}\right)^{-1}.A^T_{(0)}.\left(y - Y_{(0)}\right)$$

If $\Delta B_{(0)}$ is "equal to zero" then the estimate of B is equal to $B_{(0)}$.

(In practice, when we say "equal to zero" in this process, we really mean smaller than the approximation vector one has to define beforehand).

If $\Delta B_{(0)}$ is not "equal to zero" then the vector $B_{(0)}$ will be replaced by :

$$B_{(1)} = B_{(0)} + \Delta B_{(0)}$$

And the process will be repeated, that is, there will be another iteration with $B_{(0)}$ replaced by $B_{(1)}$ (and $A_{(0)}$ replaced by $A_{(1)}$). The iterative process will go on until the convergence at the desired level of approximation is reached.

Comments

1. It is not guaranteed that the process always converges. Sometimes it does not, some other times it is too slow (even for computers!) and some other times it converges to another limit!!
2. The above described method is the Gauss-Newton method which is the basis of many other methods. Some of those methods introduce modifications in order to obtain a faster convergence like the Marquardt method (1963), which is frequently used in fisheries research. Other methods use the second order Taylor expansion (Newton-Raphson method), looking for a better approximation. Some others, combine the two modifications.
3. These methods need the calculation of the derivatives of the functions. Some computer programs require the introduction of the mathematical expressions of the derivatives, while others use sub-routines with numerical approximations of the derivatives.
4. In fisheries research, there are methods to calculate the initial values of the parameters, for example in growth, mortality, selectivity or maturity analyses.
5. It is important to point out that the convergence of the iterative methods is faster and more likely to approach the true limit when the initial value of the vector $B_{(0)}$ is close to the real value.

7.4 ESTIMATION OF GROWTH PARAMETERS

The least squares method (non-linear regression) allows the estimation of the parameters K, L_∞ and t_o of the individual growth equations.

The starting values of K, L_∞ and t_0 for the iterative process of estimation can be obtained by simple linear regression using the following methods :

Ford-Walford (1933-1946) and Gulland and Holt (1959) Methods

The Ford-Walford and Gulland and Holt expressions, which were presented in Section 3.4, are already in their linear form, allowing the estimation of K and L_∞ with methods of simple linear regression on observed L_i and T_i. The Gulland and Holt expression allows the estimation of K and L_∞ even when the intervals of time T_i are not constant. In this case, it is convenient to re-write the expression as:

$$\Delta L/T_I = K . L_\infty - K. \ \overline{L}$$

Stamatopoulos and Caddy Method (1989)

These authors also present a method to estimate K, L_∞ and t_o (or L_o) using the simple linear regression. In this case the von Bertalanffy equation should be expressed as a linear relation of L_t against e^{-Kt}.

Consider n pairs of values t_i, L_i where t_i is the age and L_i the length of the individual i where i=1,2, ..., n.

The von Bertalanffy equation , in its general form is (as previously seen):

$$L_\infty - L_t = (L_\infty - L_a).\ e^{-K(t-ta)}$$

It can be written as:

$$L_t = L_\infty - (L_\infty - L_a).\ e^{+Kta}.\ e^{-Kt}$$

The equation has the simple linear form, y = a + bx, where:

$$y = L_t \qquad a = L_\infty \qquad b = - (L_\infty - L_a).\ e^{+Kta}$$

$$x = e^{-Kt}$$

If one takes $L_a = 0$, then $t_a = t_o$, but, if one considers $t_a = 0$, then $L_a = L_o$.

The parameters to estimate from a and b will be L_∞, t_o or L_o.

The authors propose adopting an initial value $K_{(0)}$, of K, and estimating $a_{(0)}$, $b_{(0)}$ and $r^2_{(0)}$ by simple linear regression between y (= L_t) and $x(=e^k_{(0)})$. The procedure may be repeated for several values of K, that is, $K_{(1)}$ $K_{(2)}$,.... One can then adopt the regression that results in the larger value of r^2, to which K_{max} , a_{max} and b_{max} correspond. From the values of a_{max}, b_{max} and K_{max} one can obtain the values of the remaining parameters.

One practical process towards finding K_{max} can be:

(i). To select two extreme values of K which include the required value, for example K= 0 and K=2 (for practical difficulties, use K = 0.00001 instead of K = 0).

(ii). Calculate the 10 regressions for equally-spaced values of K between those two values in regular intervals.

(iii). The corresponding 10 values of r^2 will allow one to select two new values of K which determine another interval, smaller than the one in (i), containing another maximum value of r^2.

(iv). The steps (ii) and (iii) can be repeated until an interval of values of K with the desired approximation is obtained. Generally, the steps do not need many repetitions.

7.5 ESTIMATION OF M – NATURAL MORTALITY COEFFICIENT

Several methods were proposed to estimate M, and they are based on the association of M with other biological parameters of the resource. These methods can produce approximate results.

7.5.1 RELATION OF M WITH THE LONGEVITY, t_λ

Longevity: Maximum mean age t_λ of the individuals in a non-exploited population.

Duration of the exploitable life: $t_\lambda - t_r = \lambda$ (Figure 7.1)

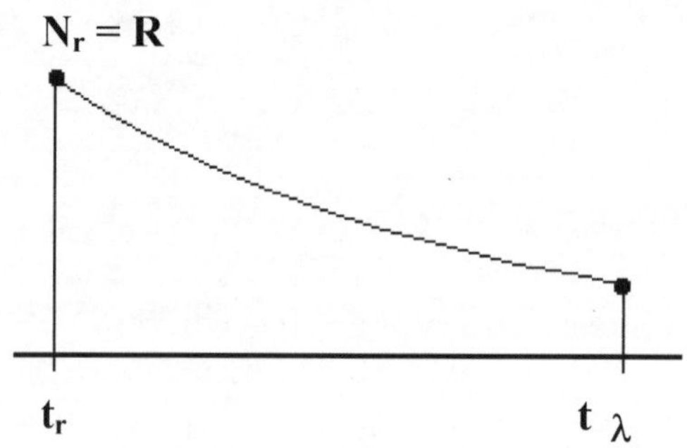

Figure 7.1 Duration of the exploitable life

Tanaka (1960) proposes "NATURAL" Survival Curves (Figure 7.2) to obtain the values of M from longevity.

A cohort practically vanishes when only a fraction, p, of the recruited individuals survives. In that case, $N_\lambda = R \cdot e^{-M \cdot \lambda}$, and it can be written:

$$p = \frac{N_\lambda}{R} = e^{-M.\lambda} \qquad \text{and so} \qquad M = -(1/\lambda).\ln p$$

Different values of the survival fraction produce different survival curves of M in function of λ.

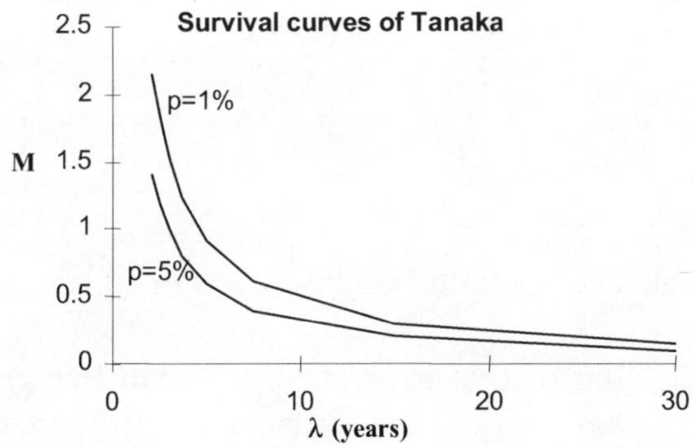

Figure 7.2 Survival curves by Tanaka

Any value of p can be chosen, for instance, p = 5%, (*i.e.* one in each twenty recruits survives until the age t_λ) as variable value of the survival curves.

7.5.2 RELATION BETWEEN M AND GROWTH

Beverton and Holt Method (1959)

Gulland (1969) mentions that Beverton and Holt verified that species with a larger mortality rate M also presented larger values of K. Looking for a simple relation between these two parameters, they concluded approximately that:

$$1 \le \frac{M}{K} \le 2 \qquad \text{for small pelagic fishes}$$

$$2 \le \frac{M}{K} \le 3 \qquad \text{for demersal fishes}$$

Pauly Method (1980)

Based on the following considerations:

1. Resources with a high mortality rate cannot have a very big maximum size;
2. In warmer waters, the metabolism is accelerated, so the individuals can grow up to a larger size and reach the maximum size faster than in colder waters.

Based on data of 175 species, Pauly adjusted multiple linear regressions of transformed values of M against the corresponding transformed values of K, L_∞ and temperature, T, and selected one that was considered to have a better adjustment, that is, the following empirical relation:

$$\ln M = -0.0152 - 0.0279 \ln L_\infty + 0.6543 \ln K + 0.463 \ln T°$$

with the parameters expressed in the following units:

$M = year^{-1}$
L_∞ = cm of total length
$K = year^{-1}$
$T°$ = surface temperature of the waters in $°C$

Pauly highlights the application of this expression to small pelagic fishes and crustaceans. The Pauly relation uses decimal logarithms to present the first coefficient different from the value -0.0152 which was given in the previous expression, written with natural logarithms.

7.5.3 RELATION BETWEEN M AND REPRODUCTION

Rikhter and Efanov Method (1976)

These authors analysed the dependency between **M** and the age of first (or 50 percent) maturity. They used data from short, mean and long life species, and suggested the following relation of **M** with the, t_{mat}, age of 1st maturity:

(Units)

$$M = \frac{1.521}{(t_{mat50\%})^{0.720}} - 0.155 \qquad \begin{pmatrix} t_{mat50\%} \rightarrow year \\ M \rightarrow year^{-1} \end{pmatrix}$$

Gundersson Method (1980)

Based on the assumption that the natural mortality rate should be related to the investment of the fish in reproduction, beyond the influence of other factors, Gundersson established several relations between *M* and those factors.

He proposed, however, the following simple empirical relation, using the Gonadosomatic Index (GSI) (estimated for mature females in the spawning period) in order to calculate M:

$$M = 4.64 \times GSI - 0.37$$

7.5.4 KNOWING THE STOCK AGE STRUCTURE, AT BEGINNING AND END OF YEAR, AND CATCHES IN NUMBER, BY AGE, DURING THE YEAR

The natural mortality coefficients M_i , at age i can be calculated from the catch, C_i , in numbers, and the survival numbers, N_i and N_{i+1} at the beginning and end of a year, by following the steps:

calculate $$E_i = \frac{C_i}{N_i - N_{i+1}}$$

calculate $$Z_i = \ln N_i - \ln N_{i+1}$$

calculate $$M_i = Z_i \cdot (1 - E_i)$$

The several values of M obtained in each age could be combined to calculate a constant value, M, for all ages.

Paloheimo Method (1961)

Let us consider the supposition that F_i is proportional to f_i for several years i, that is

$$F_i = q \cdot \frac{f_i}{T_i} \qquad \text{for } T_i = 1 \text{ year, } F_i = q \cdot f_i,$$

then: $$Z_i = q \cdot f_i + M$$

So, the linear regression between Z_i and f_i has a slope $b = q$ and an intercept $a = M$.

7.6 ESTIMATION OF Z – TOTAL MORTALITY COEFFICIENT

There are several methods of estimating the total mortality coefficient, Z, assumed to be *constant* during a certain interval of ages or years.

It is convenient to group the methods, according to the basic data, into those using ages or those using lengths.

7.6.1 METHODS USING AGE DATA

The different methods are based on the general expression of the number of survivors of a cohort, at the instant t, submitted to the total mortality, Z, during an interval of time, that is:

$$N_t = N_a . e^{-Z(t-t_a)}$$

Z is supposed to be constant in the interval of time (t_a,t_b).
Taking logarithms and re-arranging the terms, the expression will be:

$$\ln N_t = Cte - Z.t$$

where Cte is a constant ($= \ln N_a + Zt_a$).

This expression shows that the logarithm of the number of survivors *is linear* with the age, being the slope equal to -Z.

Any constant expression which does not affect the determination of Z will be referred to as Cte.

1. If Z can be considered constant inside the interval (t_a,t_b) and, having available abundance data, $\overline{N}_i$, or indices of abundance in number, $\overline{U}_i$ in several ages, i, *then,* the application of the simple linear regression allows one to estimate the total mortality coefficient Z.

In fact

$$\overline{N}_i = N_i.\frac{1-e^{-ZT_i}}{ZT_i} \qquad \text{so } \overline{N}_i = N_i. \text{ Constant}$$

and, as

$$N_i = N_a.e^{-Z(t_i-t_a)}$$

then, by substitution:

$$\overline{N}_i = \text{Cte}.e^{-Zt_i} \qquad (T_i = \text{const} = 1 \text{ year})$$

and also

$$\ln \overline{N}_i = \text{Cte} - Zt_i$$

The simple linear regression between $\ln \overline{N}_i$ and t_i allows the estimation of Z (notice that the constant, Cte is different from the previous one. In this case only the slope matters to estimate Z).

2. If ages are not at constant intervals, the expression could be approximated and expressed in terms of the $t_{centrali.}$ For T_i variable, it will be:

$$\overline{N}_i \approx N_i \,.\, e^{-ZTi/2}$$

and, as $N_i = N_a \,.\, e^{-Z.(ti-ta)}$

it will be $\overline{N}_i \approx \text{Cte}.\, e^{-Ztcentrali}$

and finally: $\ln\ \overline{N}_i \approx \text{Cte} - Z.\ t_{centrali}$

3. When using indices $\overline{U}_i$, the situation is similar because $\overline{U}_i = q.\ \overline{N}_i$, with q constant, and then, also:

$$\ln \overline{U}_i = Cte - Zt_i$$

The simple linear regression between $\ln \overline{U}_i$ and t_i allows one to estimate Z.

4. If the intervals are not constant, the expression should be modified to:

$$\ln \overline{U}_i \approx Cte - Z.\ t_{centrali}$$

Simple linear regression can be applied to obtain Z, from catches, C_i , and ages, t_i , supposing that F_i is constant.

$C_i = F_i\ \overline{N}_i\ T_i$ and so, $\ln C_i = Cte + \ln\ \overline{N}_i$ when T_i is constant. So:

$$\ln C_i = Cte - Z.\ t_i$$

5. If the intervals are not constant, the expression should be modified to:

$$\ln C_i/T_i \approx Cte - Z.\ t_{centrali}$$

6. Let V_i be the cumulative catch from t_i until the end of the life, then:

$$V_i = \Sigma\ C_k = \Sigma\ F_k\ N_{kcum},$$

Where the sum goes from the last age until age i,

As F_k and Z_k are supposed to be constant $\Sigma N_{kcum} = N_i/Z$ and so:

$$V_i = FN/Z \qquad \text{and} \qquad \ln V_i = C^{te} + \ln N_i$$

Therefore:

$$\ln V_i = Cte - Z.\ t_i$$

7. Following Beverton and Holt (1956), Z can be expressed as :

$$Z = \frac{1}{\bar{t} - t_a}$$

Then, it is possible to estimate Z from the mean age $\bar{t}$
This expression was derived, considering the interval (t_a, t_b) as (t_a, ∞).

7.6.2 METHODS USING LENGTH DATA

When one has available data by length classes instead of by age, the methods previously referred to can still be applied. For that purpose, it is convenient to define the relative age.

Using the von Bertalanffy equation one can obtain the age *t* in function of the length, as:

(the expression is written in the general form in relation to t_a and not to t_0)

$$t = t_a - (1/K).\ln[(L_\infty - L_t)/(L_\infty - L_a)]$$

or

$$t = t_a - \frac{1}{K}.\ln(1 - \frac{L_t - L_a}{L_\infty - L_a})$$

(This equation is referred to by some authors as the inverse von Bertalanffy equation).

The difference t-ta is called relative age, t^*, .

So: $t^* = -(1/K).\ln[(L_\infty - L_t)/(L_\infty - L_a)]$ or $t^* = -(1/K)\ln[1-(L_t-L_a)/(L_\infty - L_a)]$

For $t_a = t_o$, $L_a = 0$ and:

$$t^* = -\frac{1}{K}.\ln(1 - \frac{L_t}{L_\infty})$$

t^* is called a relative age because the absolute ages, t , are related to a constant age, t_a.
In this way, the duration of the interval T_i can either be calculated by the difference of the absolute ages or by the difference of the relative ages at the extremes of the interval:

$$T_i = t_{i+1} - t_i = t^*_{i+1} - t^*_i$$

Also:

$$t^*_{central\,i} = t_{central\,i} + Cte$$

$$\overline{t^*} = \overline{t} + Cte$$

So, the previous expressions still hold when the *absolute* ages are replaced by the *relative* ages:

$$\ln \overline{N}_i = Cte - Z.\ t^*_{central\,i}$$

$$\ln \overline{U}_i = Cte - Z.\ t^*_{central\,i}$$

$$\ln V_i = Cte - Z.\ t^*_i$$
$$\ln C_i/T_i = Cte - Z.\ t^*_{central\,i}$$

Finally, the expression would also be :

$$Z = \frac{1}{\overline{t^*}}$$

Beverton and Holt (1957) proved that :

$$Z = K \cdot \frac{L_\infty - \bar{L}}{\bar{L} - L_a}$$

$\bar{L}$ must be calculated as the mean of the lengths weighted with abundances (or their indices) or with the catches in numbers.

Comments

1. The application of any of these methods must be preceeded by the graphical representation of the corresponding data, in order to verify if the assumptions of the methods are acceptable or not and also to determine the adequate interval, (t_a, t_b).
2. These formulas are proved with the indications that were presented, but it is a good exercise to develop the demonstrations as they clarify the methods.
3. It is useful to estimate a constant Z, even when it is not acceptable, because it gives a general orientation about the size of the values one can expect.
4. The methods are sometimes referred to by the names of the authors. For example, the expression $\ln V_i = Cte - Z.t^*_i$ is called the Jones and van Zalinge method (1981).
5. The mean age as well as the mean length in the catch can be calculated from the following expressions:

$$\bar{t} = \frac{\Sigma(t_{centrali} \cdot C_i)}{\Sigma C_i}$$ with C_i = catch in number in the age class i

$$\bar{L} = \frac{\Sigma(L_{centrali} \cdot C_i)}{\Sigma C_i}$$

where C_i = catch in number in the length class i

$$\bar{t}^* = \frac{\Sigma(t^*_{centrali} \cdot C_i)}{\Sigma C_i}$$

with C_i = catch in number in the age class.

The relative age should be $t^* = -(1/K).\ln[(L_\infty - L_t)/(L_\infty - L_a)]$

Summary of the Methods to Estimate the Total Mortality Coefficient, Z

Assumption: Z is constant in the interval of ages, (t_a, t_b)

T Constant

$$\ln \overline{N}_i = Cte - Z \cdot t_i$$

$$\ln \overline{U}_i = Cte - Z \cdot t_i$$

$$\ln C_i = Cte - Z \cdot t_i$$

$$\ln V_i = Cte - Z \cdot t_i \qquad \left(V_i = \sum_{k=ult}^{i} C_k \right)$$

T_i variable

$$\ln \overline{N}_i = Cte - Z \cdot t_{central_i}$$

$$\ln \overline{U}_i = Cte - Z \cdot t_{central_i} \qquad t_{central_i} = \left(t_i + \frac{T_i}{2} \right)$$

$$\ln \frac{\overline{C}_i}{T_i} = Cte - Z \cdot t_{central_i}$$

$$\ln V_i = Cte - Z \cdot t_i$$

$$Z = \frac{1}{\overline{t} - t_a}$$ $(t_b = \infty)$ (Beverton and Holt equation of *Z*)

Supposition: Z is constant in the interval of lengths, **(L_a, L_b)**

Relative age $t_i^* = -\frac{1}{K} \cdot \ln\left(\frac{L_\infty - L_t}{L_\infty - La} \right)$

$$\ln N_i = Cte - Z.t_i^*$$

$$\ln \overline{N}_i = Cte - Z \cdot t^*_{central_i} \qquad t^*_{central_i} = \frac{t_i^* + t_{i+1}^*}{2}$$

$$\ln \overline{U}_i = Cte - Z \cdot t^*_{central_i}$$

$$\ln \frac{C_i}{T_i} = Cte - Z \cdot t^*_{central_i}$$ $T_i = t^*_{i+1} - t^*_i$ (Gulland and Holt equation)

$$\ln V_i = Cte - Z \cdot t_i^*$$ (Jones and van Zalinge equation)

$$Z = K \cdot \frac{L_\infty - \overline{L}}{\overline{L} - L_a}$$ $(t_b^* = \infty)$ (Beverton and Holt equation of *Z*)

7.7 ESTIMATION OF THE PARAMETERS OF THE STOCK-RECRUITMENT (S-R) RELATION

The least squares method (non-linear model) can be used to estimate the parameters, α and k, of any of the S-R models.

The initial values of the Beverton and Holt model (1957) can be obtained by re-writing the equation as:

$$(R/S)^{-1} \text{ or } \frac{S}{R} = \frac{1}{\alpha} + \frac{1}{\alpha k}.S$$

and estimating the simple linear regression between y (= S/R) and x (=S) which will give the estimations of $1/\alpha$ and $1/(\alpha k)$. From these values, it will then be possible to estimate the parameters α and k. These values can be considered as the initial values in the application of the non-linear model.

In the Ricker model (1954) the parameters can be obtained by re-writing the equation as:

$$\ln\frac{R}{S} = \ln\alpha - \frac{1}{k}.S$$

and applying the simple linear regression between y (= ln R/S) and x (=S) to estimate $\ln\alpha$ and (-1/k). From these values, it will be possible to estimate the parameters (α and k) of the model, which can be considered as the initial values in the application of the non-linear model.

It is useful to represent the graph of y against x in order to verify if the marked points are adjustable to a straight line before applying the linear regression in any of these models.

In the models with the flexible parameter, c, like for example, the Deriso model (1980), the equation can be re-written as:

$$\left(\frac{R}{S}\right)^c = \alpha^c - c.\alpha^c.\frac{S}{k}$$

For a given value of c the linear regression between y (= $(R/S)^c$) and x (=S) allows the estimation of the parameters α and k.

One can try several values of c to verify which one will have a better adjustment with the line y against x; for example, values of c between -1 and 1.

The values thus obtained for α, k and c, can be considered as initial values in the application of the iterative method, to estimate the parameters α , k and c of the non-linear Deriso model.

7.8 ESTIMATION OF THE MATRIX [F] AND OF THE MATRIX [N] – COHORT ANALYSIS – AC and LCA

7.8.1 COHORT ANALYSIS BY AGE- (AC)

The cohort analysis is a method to estimate the fishing mortality coefficients, F_i, and the number of survivors, N_i, at the beginning of each age, from the annual structures of the stock catches, in number, over a period of years.

More specifically, consider a stock where the following is known:

Data

age, i, where i = 1,2,...,k
year, j, where j = 1,2,...,n

Matrix of catches [C] with
$C_{i,j}$ = Annual catch, in number, of the individuals with the age i and during the year j

Matrix of natural mortality [M] with
$M_{i,j}$ = natural mortality coefficient, at the age i and in the year j.

Vector [T] where
T_i = Size of the age interval i (in general, T_i=T=1 year)

Objective

To estimate

matrix [F]

and

matrix [N].

In the resolution of this problem, it is convenient to consider these estimations separately; one interval of age i (part 1); all the ages during the life of a cohort (part 2); and finally, all the ages and years (part 3).

PART 1 (INTERVAL T_I)

Consider that the following characteristics of a cohort, in an interval T_i are known :

C_i = Catch in number

M_i = Natural mortality coefficient

T_i = Size of the interval

Adopting a value of F_i, it is then possible to estimate the number of survivors at the beginning, N_i , and at the end, N_{i+1} , of the interval.

In fact, from the expression:

$$C_i = \frac{F_i}{F_i + M_i}.N_i.(1 - e^{-(F_i+M_i).T_i})$$

one can calculate N_i which is the only unknown variable in the expression.

To calculate N_{i+1} one can use the expression $N_{i+1} = N_i.e^{-(F_i+M_i).T_i}$ where the values N_i, F_i and M_i were previously obtained.

PART 2 (DURING THE LIFE)

Suppose now that the catches C_i of each age i, of a cohort during its life, the values of M_i and the sizes of the interval T_i are known.

Adopting a certain value, F_{final}, for the Fishing Mortality Coefficient in the last class of ages, it is possible, as mentioned in part 1, to estimate all the parameters (related to numbers) *in that last age group*. In this way, one will know the number of survivors at the beginning and end of the last age .

The number at the beginning of that last class of ages, is also the number N_{last} at the end of the previous class, that is, N_{final} is the initial number of survivors of the class before last.

Using the C_i expression, resulting from the combination of the two expressions above :

$$C_i = \frac{F_i}{F_i + M_i}.N_{final}.(e^{+(F_i+M_i).T_i} - 1)$$

one can estimate F_i in the previous class, which is the only unknown variable in the expression. The estimation may require iterative methods or trial and error methods.

Finally, to estimate the number N_i of survivors at the beginning of the class i, the following expression can be used :

$$N_i = N_{final}.e^{(F_i+M_i).T_i}$$

Repeating this process for all previous classes, one will successively obtain the parameters in all ages, until the first age.

In the case of a completely caught cohort, the number at the end of the last class is zero and the catch C has to be expressed as :

$$C_{final} = \frac{F_{final}}{(F_{final} + M_i)}.N_{final}$$

Pope Method

Pope (1972) presented a simple method to estimate the number of survivors at the beginning of each age of the cohort life, starting from the last age.

It is enough to apply successively in a backward way, the expression:

$$N_i \approx (N_{i+1}\, e^{MT/2} + C_i).e^{MT/2}$$

Pope indicates that the approximation is good when $MT \leq 0.6$

Pope's expression is obtained, supposing that the catch is made exactly at the central point of the interval T_i (Figure 7.3).

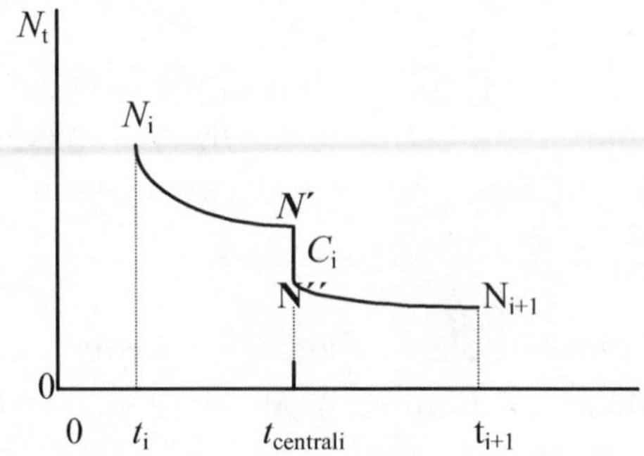

Figure 7.3 Number of survivors during the interval $T_i = t_{i+1} - t_i$ with the catch extracted at the central point of the interval

Proceeding from the end to the beginning one calculates successively:

$$N'' = N_{i+1}e^{+MTi/2}$$

$$N' = N'' + C_i$$

$$\mathbf{N_i} = N'.e^{+MT/2}$$

substituting N' by N''+C_i, the expression will be:

$$\mathbf{N_i} = (N'' + C_i).e^{MT/2}$$

Finally, substituting N'' by $N_{i+1}.e^{+MTi/2}$, it will be:

$$N_i \approx \left(N_{i+1}e^{MT/2} + C_i\right).e^{MT/2}$$

Part 3 (period of years)

Let us suppose now that the Catch matrix [C], the natural mortality [M] matrix and the vector size of the intervals [T], are known for a period of years.

Let us also assume that the values of F in the last age of all the years represented in the matrices and the values of F of all the ages of the last year were adopted. These values will be designated by $F_{terminal}$ (Figure 7.4)

	Years				
Ages	2000	2001	2002	2003	
1	C	C	C	C	$F_{terminal}$
2	C	C	C	C	$F_{terminal}$
3	C	C	C	C	$F_{terminal}$
	$F_{terminal}$	$F_{terminal}$	$F_{terminal}$	$F_{terminal}$	

Figure 7.4 Matrix of catch, [C], with $F_{terminal}$ in the last line and in the last column of the matrix C. The shadowed zones exemplify the catches of a cohort

Notice that in this matrix the elements of the diagonal correspond to values of the same cohort, because one element of a certain age and a certain year will be followed, in the diagonal, by the element that is a year older.

From parts 1 and 2 it will then be possible to estimate successively Fs and Ns for all the cohorts present in the catch matrix.

Comments

1. The values of $M_{i,j}$ are considered constant and equal to M, when there is no information to adopt other values.

2. When data is referred to ages, the values T_i will be equal to 1 year.

3. The last age group of each year is, sometimes grouped ages(+). The corresponding catches are composed of individuals caught during those years, with several ages. So, the cumulative values do not belong to the same cohorts, but are survivors of several previous cohorts with different recruitments and submitted to different fishing patterns. It would not be appropriate to use the catch of a group (+) and to apply cohort analysis. Despite this fact, the group (+) is important in order to calculate the annual totals of the catches in weight, Y, of total biomasses, B, and the spawning stock biomass. So, it is usual to start with the cohort analysis on the age immediately before the group (+) and use the group (+) only to calculate the annuals Y, B and (SP). The value of F in that group (+) in each year, can be estimated as being the same fishing mortality coefficient as the previous age or, in some cases, as being a reasonable value in relation to the values of F_i in the year that is being considered.

4. A difficulty in the technical application appears when the number of ages is small or when the years are few. In fact, in those cases, the cohorts have few age classes represented in the Matrix [C] and the estimations will be very dependent on the adopted values of $F_{terminals}$.

5. The cohort analysis (CA) has also been designated as: VPA (Virtual Population Analysis), Derzhavin method, Murphy method, Gulland method, Pope method, Sequential Analysis, etc. Sometimes, CA is referred to when the Pope formula and the VPA are used in other cases. Megrey (1989) presents a very complete revision about the cohort analyses.

6. It is also possible to estimate the remaining parameters in an age *i*, related to numbers, that is, N_{cumi}, $\overline{N}_i$, D_i, Z_i and E_i. When the information on initial individual or mean weights matrices [w] or [$\overline{w}$] are available, one can also calculate the matrices of annual catch in weight [Y], of biomasses at the beginning of the years, [B], and of mean biomasses during the years [$\overline{B}$]. If one has information on maturity ogives in each year, for example at the beginning of the year, spawning biomasses [SP] can also be calculated. Usually, only the total catches Y, the stock biomasses (total and spawning) at the beginning and the mean biomasses of the stock (total and spawning) in each year are estimated.

7. The elements on the first line of the matrix [N] can be considered estimates of the recruitment to the fishery in each year.

8. The fact that the $F_{terminals}$ are adopted and that these values have influence on the resulting matrix [F] and matrix [N], forces the selection of values of $F_{terminals}$ to be near the real ones. The agreement between the estimations of the parameters mentioned in the points 6. and 7. and other independent data or indices (for example, estimations by acoustic methods of recruitment or biomasses, estimations of abundance indices or cpue´s, of fishing efforts, etc) must be analysed.

9. The hypothesis that the exploitation pattern is constant from year to year, means that the fishing level and the exploitation pattern can be separated, or $F_{sepi} = F_j \times s_i$. This hypothesis can be tested based on the matrix [F] obtained from the cohort analysis.

   It is usual to call this separation VPA-Separable (SVPA).

   We have $\sum\limits_{i} F_{i,j} = F_{tot_j}$

   and $\sum\limits_{j} F_{i,j} = s_{tot_i}$

   and $\sum\limits_{i,j} F_{i,j} = F_{tot}$

   Then, if $F_{ij} = F_j.s_i$ one can prove that $F_j.s_i = (F_{tot_j}.s_{tot_i})/F_{tot}$.

   If the estimated values of F_{ij} are the same as the previous $Fsep_{ij} = F_j.s_i$ then the hypothesis is verified. This comparison can be carried out in two different ways, the simplest is to calculate the quotients $Fsep_{ij}/F_{ij}$. If the hypothesis is true this quotient is equal to one. If the hypothesis is not verified it is always possible to consider other hypotheses with the annual vector [s] constant in some years only, mainly the last years.

10. It is usual to consider an interval of ages, where it can be assumed that the individuals caught are "completely recruited". In that case, the interval of ages corresponds to exploitation pattern constant (for the remaining ages, not completely recruited, the exploitation pattern should be smaller). For that interval of ages, the means of the values

of $F_{i,j}$ in each year are then calculated. Those means, $\overline{F}_j$, are considered as fishing levels in the respective years. The exploitation pattern in each cell, would then be the ratio $F_{i,j}$ / $\overline{F}_j$. The s_i, for the period of years considered, can be taken as the mean of the relative pattern of exploitation calculated before. Alternatively, they can also be taken as referring to s_i of an age chosen for reference.

7.8.2 LENGTH COHORT ANALYSIS - (LCA)

The technique of the cohort ansalysis, applied to the structure of the catches of a cohort during its life, can be made with non constant intervals of time, T_i,. This means that the length classes structure of the catches of a cohort during its life, can also be analysed.

The methods of analysis of the cohort in those cases is called the LCA (**L**ength **C**ohort **A**nalysis). The same techniques; Pope method, iterative method, etc, of the CA for the ages, can be applied to the LCA analysis (the intervals T_i's can be calculated from the relative ages).

One way to apply the LCA to the length annual catch compositions, will be: to group the catches of length classes belonging to the same age interval in each year. The technique CA can then be applied directly to the resulting age composition of the catches by age of the matrix [C]. This technique is known as "slicing" the length compositions. To "slice", one usually inverts the von Bertalanffy length growth equation and estimates the age t_i for each length L_i (sometimes using the relative ages t^*_i) (Figure 7.5). It is possible that when grouping the length classes of the respective age interval, there are length classes composed by elements that belong to two consecutive age groups. In these cases, it will be necessary to "break" the catch of these extreme classes into two parts and distribute them to each of those ages. In the example of Figure 7.5, the catches of the length class (24-26] belong to age 0 and to age 1 . So, it is necessary to distribute that catch to the two ages. One simple method is to attribute to age 0 the fraction (1.00 - 0.98)/(1.06 - 0.98) = 0.25 of the annual catch of that length class and to age 1 the fraction (1.06 - 1.00)/(1.06 - 0.98) = 0.75. The method may not be the most appropriate one, because it is based on the assumption that, in the length classes, the distribution of the individuals by length is uniform. So, it is necessary to use the smallest possible interval of length classes, when applying this distribution technique.

Another way to do the length cohort analysis is to use the catches in the length classes of the same age group. It is possible to follow the cohorts in the matrix [C], through the length classes belonging to a same age, in a certain year, with the length classes of the next age, in the following year, etc. In this way, the different cohorts existing in the matrix will be separated and the evolution of each one of them will be by length classes, not by age (see Figure 7.5).

Group Age	Relative age	Length Classes	Years 2000	2001	2002	2003
	1.03	20-	**41**	30	17	49
0	1.54	22-	**400**	292	166	472
	1.98	24-	**952**	699	400	1127
	2.06	26-	1766	**1317**	757	2108
1	2.30	28-	2222	**1702**	985	2688
	2.74	30-	2357	**1872**	1093	2902
	2.88	32-	2175	**1091**	1067	2739
	3.00	34-	1817	948	**1416**	1445
	3.42	36-	1529	812	**1270**	1250
2	3.64	38-	1251	684	**980**	1053
	3.83	40-	1003	560	**702**	710
	3.96	42-	787	290	**310**	558
3	4.01	44-	595	226	179	**834**
	4.25	46-	168	70	71	**112**

Cohort of the year 2000

Figure 7.5 Example of a matrix [C] with the catches of the cohort shadowed, written in bold, recruited at year 2000, "sliced" by length classes,

The LCA R. Jones method (1961), of analysing a length composition during the life of *a* cohort can then be applied. The different values of T_i are calculated as $T_i = t_{i+1}^* - t_i^*$, where t_i^* e t_{i+1}^* are the relative ages corresponding to the extremes of the length interval i. The vector [N] can also be obtained as the number of initial survivors in each length class of the cohort, and in each age class.

Comments on cohort analyses

1. Certain models, called *integrated models*, combine all the available information (catches, data collected on research cruises, effort and cpue data, etc) with the matrix [**C**], and integrate in a unique model, in order to optimize the previously defined criterium function. A model integrating CA and the hypothesis of constant exploitation pattern was developed and called SVPA, separable VPA, because the Fishing level and Exploitation pattern are "separable".

2. Fry (1949) considered the cumulative catches of a cohort by age during its life, from the end to the beginning, as an image of the number of survivors at the beginning of each age (which the author designated as "virtual population"):

$$\sum_{k=final}^{i} C_k = V_i = N_i \text{virtual}$$

In the fishery that Fry studied, M was practically equal to zero.

If M is different from zero it can also be said that the number N_i of survivors at the beginning of the interval **i** will be :

$$N_i = \sum_{k=final}^{i} D_k$$

where D_k represents the number of total deaths at the interval k.

Adopting the initial values, $E_{k(0)}$, for the exploitation rates, E , in all the classes, one can calculate the total deaths:

$$D_{k(0)} = C_k / E_{k(0)}.$$

$N_{i(0)}$ can be calculated as the cumulative total deaths from the last class up to the ith class, that is:

$$N_{i(0)} = \sum_{k=ult}^{i} D_{k(o)}$$

Then the expression will be:

$$Z_{i(1)}.T_i = \ln(N_{i+1(0)} / N_{i(0)})$$

and:

$$F_{i(1)}.T_i = E_{i(0)}.Z_{i(1)}.T_i$$

Comparing $E_{i(1)}$ with $E_{i(0)}$, the new values of E will be:

$$E_{i(1)} = F_{i(1)}.T_i / (F_{i(1)}.T_i + M_i.T_i)$$

One can then estimate the values of E with the desired approximation by an iterative method, repeating the five calculations (of D_i, N_i, Z_iT_i, F_iT_i and E_i,) using $E_{i(1)}$ instead of $E_{i(0)}$.

In the last class, the number, N_{last}, can be taken as equal to the number of deaths, D_{last}, and in this case, N_{last} will be calculated as :

$$N_{last} = D_{last} = C_{last} / E_{last}$$

3. Finally, the results of CA and of LCA give a perspective view of the stock in the previous years. That information is useful for the short and long–term projections. Usually, data concerning the catches is not available for the year in which the assessment is done and so it is necessary to project the catches and the biomasses to the beginning of the present year before calculating the short–term projection.

4. When the relative ages are calculated, it is usual to adopt zero as the age t_a corresponding to the value of L_a , taken as the lower limit of the first length class represented in the catches.

CHAPTER 8 – EXERCISES

8.1 MATHEMATICAL REVISION

1. Calculate:

A) 10^4 $\quad 16^{-\frac{1}{2}}$ $\quad 8427^0$ $\quad 0.01^{0.5}$

B) $\frac{4^3}{4^5}$ $\quad \left(3^{-4}\right)^4$ $\quad 5^2+4^2$ $\quad 2^2\times 2^5$

C) $\log 1000$ $\quad \log 0.01$ $\quad \log\left(\frac{10^4}{10^{-3}}\right)$

D) $\ln e$ $\quad \ln\frac{1}{e}$ $\quad \ln e^{-5}$ $\quad e^{\ln e}$

E) $\frac{0}{0}$ $\quad \frac{\mathrm{A}}{0}$ $\quad \frac{0}{\mathrm{A}}$ $\quad \frac{\infty}{\infty}$ $\quad \frac{\mathrm{A}}{\infty}$ $\quad \frac{\infty}{\mathrm{A}}$ $\quad \frac{0}{\infty}$ $\quad \frac{\infty}{0}$

2. Verify that

a) $a = e^{\ln a}$

b) $a = 10^{\log a}$

c) $\frac{e^x-1}{x} \approx 1$ for $-0.01 < x < +0.01$

d) $\frac{e^x-1}{x} \approx e^{\frac{x}{2}}$ for $-0.5 < x < +0.5$

3. Solve the following expressions applying natural logarithms to both members of the equality:

a) $y = a\cdot x^5$

b) $y = a\cdot e^{-b\cdot(x+2\cdot c)}$

c) $y - a = b\cdot e^{-c\cdot(x-b)}$

Note: *a*, *b*, e *c* are constants; *e* is the basis of natural logarithms $(e = 2.7183..)$; *x* and *y* are variables.

4. Determine the value of *x* in the following expressions:

a) $e^{-x} = 5.2$

b) $10^x = 5.5$

c) $y - a = b\cdot e^{c\cdot(x-b)}$

5. Calculate the derivatives of the following expressions:

a)	$y=13$	g)	$y=5^x$	m)	$y=(4+2x)^3$
b)	$y=3\text{-}8x$	h)	$y=e^{-3.x}$	n)	$y=(x-6)^2$
c)	$y=x^5$	i)	$y=\ln x$	o)	$y=a.(3-e^{-b.x})^3$
d)	$y=x^{2/7}$	j)	$y=\ln(5x+4)$	p)	$y=(4x+3).(e^x-4)$
e)	$y=x^{-3}$	k)	$y=1/x$		
f)	$y=e^{3.x}$	l)	$y=(2+4x)/(3-x)$		

6. Calculate the indefinite integrals of the following functions:

a) $f(x)=0$

b) $f(x)=5.34$

c) $f(x)=x^6$

d) $f(x)=1+3\cdot x$

e) $f(x)=4\cdot x^{-3}$

f) $f(x)=\frac{1}{x}$

g) $f(x)=\frac{-5}{2-5\cdot x}$

h) $f(x)=\frac{1+2\cdot x}{40+x+x^2}$

i) $f(x)=e^x$

j) $f(x)=e^{0.2\cdot x}$

k) $f(x)=e^{-0.5\cdot x}$

l) $f(x)=3\cdot e^{2\cdot x+1}$

m) $f(x)=x\cdot e^x$

n) $f(x)=\ln x$

o) $f(x)=x\cdot \ln x$

7. Calculate the area under the function

a) $f(x)=2+5x$ between $x=1$ and $x=4$

b) $f(x)=e^{3.x}$ between $x=0$ and $x=1$

c) $f(x)=\frac{2}{5+2x}$ between $x=\frac{1}{2}$ and $x=\frac{3}{2}$

d) $f(x)=1+3x$ between $x=-2$ and $x=2$

8. Calculate the value of $y_{cumulative}$ with

a) $y=e^{-2x}$ between $x=0$ and $x=0.8$

b) $f(x)=\frac{2}{1+2x}$ between $x=0$ and $x=2$

c) $f(x)=2.x^3$ between $x=0$ and $x=1$

9. Calculate the Mean Value of y with

a) $y = 3 \cdot e^{-7x}$ between $x = 0$ and $x = 1$

b) $y = 4 \cdot \left(1 - e^{-0.2 \cdot x}\right)$ between $x = 1$ and $x = 3$

c) $y = 2 - x$ between $x = 0$ and $x = 1.2$

10. Calculate the integral of

a) $f(x) = 2 \cdot e^{-0.5 \cdot x}$ with the initial condition $x = 1 \Rightarrow F(x) = 4$ where $F(x) = \int f(x) \cdot dx$

b) $f(x) = \frac{3}{x}$ with the initial condition $F(1) = 2$

c) $\frac{dy}{dx} = 0.2 \cdot y$ with the initial condition $x = 0 \Rightarrow y = 10$

d) $\frac{dy}{dx} = \frac{3}{1 + 3x}$ with the initial condition $x = 0 \Rightarrow y = 0$

8.2 RATES (2.2)

Consider the function, $y = 40 - 35.e^{-0.2x}$ at the interval (0,10)

1. Calculate:

a) The values of y for $x = 0,1,2,3,4,5,6,7,8,9,10$;

b) Represent graphically the function y at the interval $(0,10)$ of x;

c) The variation, Δy, corresponding to the interval $(1,2)$ of x;

d) The absolute mean rate of variation of y, *amr(y)*, at the intervals (1,7), (2,5), (5,6) and (8,9) of x;

e) The absolute instantaneous rate of variation of y, *air(y)*, at the points x=3 and x=4;

f) Calculate the relative mean rate of variation of y, *r.m.r.(y)*, at the interval (8,9) in relation to the value of y corresponding to the initial point, to the final point and to the central point of that interval;

g) Calculate a relative instantaneous rate of variation of y, *r.i.r.(y)* at the central point of the interval (8,9).

2. Calculate the *air(y)* of the following functions:

a) $y = 1 + 10x$

b) $y = x^3 - 2x + 3$

c) $y = e^x$

d) $y = \ln x$

3. Calculate the *rir(y)* of the following functions:

a) $y = 4 + x$

b) $y = e^x$

c) $y = 6 \cdot e^{2x}$

d) $y = a \cdot x$ with a = constant

4. Calculate the *air* of the *air(y)* of $y = 3x^2 - 4x - 12$

5. Given the function, $y = 3 \cdot e^{-1.8x}$ verify that rir(y)=air(ln y)

8.3 SIMPLE LINEAR MODEL (2.3)

Consider a model that relates the characteristic $\boldsymbol{y}$ with time t, where the basic assumption is:

air(y)= -3, for $0 < t < \infty$

Adopt the initial conditionl: for $t = 0$, y=30

1. How would you designate this model?
   Write the general expression for the value of the characteristic $\boldsymbol{y}$ at the instant t;

2. Calculate the value of y when t = 0,1,2,3,4,5,6 and draw the graph of y against t.

3. Considering the interval of time, Δt, from $t = 2$ to $t = 4$

   a) Calculate the variation of y during the mentioned interval Δt;

   b) Calculate the central value of y in the interval Δt;

   c) Calculate the cumulative value of y in that interval, y_{cum};

   d) Calculate the mean value, $\bar{y}$, of y, in the interval Δt;

e) Calculate the simple arithmetic mean of y in the interval Δt;

f) Verify that the arithmetic mean of y is equal to the mean value, $\bar{y}$, and equal to the central value, $y_{central}$, of y in that interval.

g) Verify that in the linear model, the amr(y) = air(y) = constant. To do that, calculate, for the above mentioned interval, Δt, the amr(y) and the air(y) and compare the results.

Repeat exercise 3. considering the interval from $t = 0$ to $t = 10$.

8.4 EXPONENTIAL MODEL (2.4)

Consider a model that relates the characteristic y with time t, through the following basic assumption:

$$rir(y) = -0.4 \qquad \text{for } 0 < t < \infty$$

Adopt the initial condition : for $t = 0$, $y = 100$

1. Write the general expression for the value of the characteristic y at the instant t;

   a) Calculate the value of y at the instants t =1,2,3,4,5,6.

   b) Represent, graphically, the values of y calculated above, against the corresponding values of t.

   c) Represent, graphically, the values of lny against the given values of t.

2. Considering the interval of time $\Delta t = (3,6)$

   a) Calculate the variation of y, Δy, during the interval Δt.

   b) Calculate $y_{central}$ in the interval Δt.

   c) Calculate the value of y_{cum} in the interval Δt.

   d) Calculate $\bar{y}$ in the interval Δt.

   e) Show that the geometric mean of the values of y for $t = 3$ y $t = 6$ is equal to $y_{central}$ and approximately equal to $\bar{y}$ in that interval.

   f) Show that, in that interval, $rmr(y)_{relative to \bar{y}_i} = rir(y) = -0.4$

3. Consider the interval of time from $t = 0$ to $t = 10$. Repeat the calculations of questions 2 item a), c) and d) for this interval.

8.5 COHORT – EVOLUTION IN NUMBER (3.2)

GROUP I

From the stock of megrim, *Lepidorhombus whiffiagonis*, from Divisions VIIIc and IXa of the ICES, the Assessment Working Group of ICES (ICES, 1997a) estimated that the fish recruit to the exploitable phase, at the beginning of age 1 year, and that in 1996 the instantaneous rate of total mortality during the exploitable phase was 0.7 $year^{-1}$.

Consider 1000 individuals of a cohort of megrim, recruited to the exploited phase, which starts at the beginning of age 1 and finishes at the end of age 7 years.

1.

a) What is the value of the rir of the variation of N_t in this interval ?

b) What is the value of the rir of the mortality of N_t in this interval ?

c) Calculate the annual rate of survival during the interval.

d) Calculate the annual rate of mortality during the interval.

2.

a) Calculate the number of survivors at the beginning of each age of that interval.

b) Calculate the number of survivors at the end of 7 years of age.

c) Draw the graph of the number of survivors in each age of that interval.

d) Calculate the number of deaths in each age of the interval.

e) Calculate the number of deaths during all the exploitable phase.

f) Determine the percentage of the initial number of 3 year-olds that survive until the beginning of their 6th year.

g) Determine the percentage of the initial number of 3 year-olds that die before the beginning of their 6th year.

h) Calculate the mean number of survivors during each age of the given interval.

i) Calculate the cumulative number of survivors during ages 3 to 5.

j) Calculate the mean number of survivors between the beginnings of ages 3 and 6.

GROUP II

The Working Group of ICES which evaluated the stock of iberic sardine, *Sardina pilchardus*, estimated the mortality rates in each age of 1995 (ICES, 1997b), presented in the following table :

Age Group	0	1	2	3	4	5	6
Annual Rate of Mortality	0.36	0.43	0.54	0.63	0.66	0.68	0.72

Suppose that these rates correspond to a cohort.

1. As previously seen, there may be several types of rates (ex: amr, rmr, air and rir and the relative rates were referred to several values of the characteristics). What type of rate is the annual rate of mortality?
2. Calculate the survival rate in each age class.
3. Calculate the total mortality coefficient for each age class.
4. Calculate the survival rate between the beginning of age 1 and the end of age 4.
5. Calculate the annual mean rate of survival in the same interval of ages.

GROUP III

Consider a cohort of a certain species for which the number of survivors at the beginning of age 2 years is 4325, while the number of survivors at the end of age 2 years is 2040.

1. Calculate the mean number of 2 year-old individuals and the number of individuals at the age of 2.5 years. Compare the results.

2. If the annual rate of mortality of this cohort during the ages 3 and 4 is 70 and 60 percent respectively, calculate the percentage of the initial number of individuals at age 3 that will survive until the end of age 4.

Give the relation between the survival rate during the period that covers the ages 3 and 4 years and the annual survival rates of ages 3 and 4 years.

8.6 COHORT – CATCH IN NUMBER (3.3)

GROUP I

According to the Assessment Working Group of ICES (ICES, 1996a) the relative instantaneous rates of total and natural mortality for the age of 3 years, of the stock of blue whiting, *Micromesistius poutassou*, in 1995, were estimated as being, respectively, $Z_3 = 0.4$ year^{-1} and $M_3 = 0.2$ year^{-1}. In that year, the number of survivors at the beginning of age 3 year was 2600 million individuals.

1. Calculate, for the cohort of 1992 and for the age of 3 years :

a) The total annual rates of survival and of mortality.

b) The relative instantaneous rate of fishing mortality.

c) The exploitation rate.

d) The number of deaths during the age.

e) The mean number of survivors during the age.

f) The total catch in number of 3 years old individuals.

g) The number of survivors at the end of the age.

GROUP II

The 4 year-old age group of the stock (Div. ICES VIIe-h) of whiting, *Merlangius merlangus merlangus*, is simultaneously exploited by the crustaceans trawl fleet and demersal fish trawl fleet.

The Working Group of ICES that evaluates this stock estimated (ICES, 1996a) that, in 1995, the total instantaneous mortality rate of that age (4 years) was $Z = 1.3$ year^{-1}. Suppose, for this exercise, that the instantaneous rate of fishing mortality caused by the crustaceans trawl fleet was $Fc = 0.5$ year^{-1}, while the corresponding value for the demersal fish trawl fleet was $F_f = 0.6$ year^{-1}. The Group also considered the natural instantaneous mortality rate, $M = 0.2$ year^{-1}.

1. In 1995 17.66 million individuals recruited at 4 years of age.

a) Calculate the total number of deaths during that age.

b) Calculate the mean number of survivors during the age.

c) Calculate the exploitation rate of each fleet.

d) Calculate the total exploitation rate.

e) Calculate the catch in number by each fleet and the total catch in number.

f) Calculate the number of survivors at the end of the age. (The solutions to the questions can be done in any order).

GROUP III

For this exercise, suppose that in 1990, the mean number of survivors of the cohort of a stock of anchovy, *Engraulis encrasicholus*, during the period of 2 years of age was calculated as about 50 million individuals. During 1992, 70 million individuals were caught, from which, 40 percent were caught by the national fleet, and it is estimated that 80 million died of natural causes.

1. For this age and this cohort:
   a) Calculate the total exploitation rate and the exploitation rate of the national and foreign fleets.
   b) Calculate the total, natural and fishing mortality coefficients.
   c) Calculate the fishing instantaneous mortality rates caused by the national fleet and by the foreign fleet.
   d) Calculate the number of survivors at the beginning and at the end of the age. (The solutions to the questions can be done in any order).

GROUP IV

According to the Assessment Working Group of ICES (ICES, 1997a) the fishing mortality coefficients applied to the 1976 cohort of the stock of common sole, *Solea vulgaris*, of the Celtic Sea were estimated in each age from 2 to 8 years (following table). The natural mortality coefficient for this stock is considered constant and equal to 0.1 $year^{-1}$. It was estimated that, at the beginning of age 6 years there were 1112 million survivors of this cohort.

Age	2	3	4	5	6	7	8
F_i	0.07	0.22	0.33	0.41	0.45	0.41	0.74

1. Calculate the numbers of survivors of this cohort at the beginning of each of the above mentioned ages.
2. Calculate the number of deaths in each age referred in the table.
3. Calculate the exploitation rates of this cohort in each age.
4. Calculate the mean numbers of survivors during each of the above ages.
5. Calculate the catches, in numbers, extracted from this cohort during each of the ages mentioned above, using two different methods.

8.7 INDIVIDUAL GROWTH IN LENGTH AND WEIGHT (3.4)

GROUP I

The parameters of the von Bertalanffy length growth equation of the stock of European anglerfish (Div. VIIIc and IXa of ICES), *Lophius budegassa*, were estimated as (Duarte *et al.*, 1997):

Asymptotic length = 101.69 cm
Coefficient of growth in length = 0.08 $year^{-1}$
Theoretical age, when the length is zero = -0.2 year

1. Calculate the theoretical length corresponding to the age 3.84 years.
2. Calculate the length at the beginning of the ages 1 to 12 years.
3. Calculate, for each of the above mentioned ages, the central length.
4. Represent, graphically, the Bertalanffy curve of growth in length for this stock.

GROUP II

Using the growth parameters given in Group I:

1. Calculate the length that corresponds to each age interval between 1 and 12 years as being the simple arithmetic mean of the length at the beginning and at the end of each class.
2. Calculate the mean length in each age for the same interval from 1 to 12 years accordingly to the von Bertalanffy model.
3. Compare the lengths obtained in 1) with those obtained in 2) and with the central values in each age of the interval, calculated in Group I-3.

GROUP III

The data presented in the following table represents the mean length (cm) by age (years) obtained from direct age reading of individuals of the stock of European anglerfish, *Lophius budegassa,* (Div. VIIIc and IXa).

t	L_t (cm)	T	L_t(cm)
1	9.2	7	44.4
2	16.5	8	49.0
3	22.9	9	52.3
4	28.8	10	55.0
5	34.7	11	60.8
6	38.6	12	63.4

Based on this data the parameters of the growth equation were estimated according to the Gompertz model as being :

Gompertz: L_∞ = 73.7 cm; k = 0.22 $year^{-1}$ t* = - 2.76 year

(t* is the age corresponding to L=1 cm)

1. Represent, graphically, the observed values.
2. Calculate, for the interval 1-12 years, the values of the length at the beginning of each age, according to the Bertalanffy growth model and draw the corresponding growth curve.
3. Calculate, for the interval 1-12 years, the values of the length at the beginning of each age, according to the Gompertz growth model and draw the corresponding growth curve. Determine the inflection point of the curve.
4. Say which growth model you consider more appropriate for this case and justify your answer.

GROUP IV

The data presented on the following table concern the stock of European anglerfish *Lophius budegassa* (Div. VIIIc and IXa).

Table of individual weights by length class, taken from the samples of European anglerfish, *Lophius budegassa* collected by IEO and IPIMAR in 1994.

Li (cm)	W_{meani} (g)	n	Li (cm)	W_{meani} (g)	n
20-	129	3	50-	1685	28
22-	163	2	52-	1896	30
24-	219	4	54-	2107	24
26-	265	14	56-	2345	41
28-	320	8	58-	2569	41
30-	397	10	60-	2848	32
32-	486	9	62-	3126	35
34-	545	57	64-	3407	28
36-	664	60	66-	3700	19
38-	773	61	68-	4056	23
40-	890	58	70-	4411	17
42-	1027	64	72-	4764	13
44-	1122	56	74-	5203	8
46-	1334	50	76-	5587	4
48-	1503	37	78-	5982	3

The mean of the observed weights and the number (n) of individuals, are indicated for each length class.

Based on this data, the parameters of the weight-length relation were estimated for this stock as :

$a = 0.021$
$b = 2.88$

1. Calculate the theoretical weight for each length class.
2. On a graph, plot the observed and the theoretical weights against the length classes.

3. Suppose you want to use the weight-length relation, with b=3 (the constant estimated for this relation being a=0.013). Calculate, for this case, the theoretical weights for each length class. Compare these values with the theoretical weights estimated in 1).
4. Using the results obtained up to now, write the Bertalanffy growth equation, in weight, for this stock.

8.8 COHORT DURING ALL LIFE – BIOMASS AND CATCH IN WEIGHT (3.6)

GROUP I

The recruitment to the exploitable phase of horse-mackerel, *Trachurus trachurus*, distributed in the Ibero-Atlantic waters (Div. VIIIc and IXa) occurs at age 1. In order to make the calculations, let us consider the exploitable phase between the ages 1 and 10 years and the recruitment of a cohort equal to 1 000 individuals.

The parameters of the von Bertalanffy equation were estimated by the Working Group of ICES (ICES, 1998a), using the mean lengths, at age of the catch, as being:

$L_\infty = 34.46$ cm
$K = 0.225$ year^{-1}
$t_o = -1.66$ year

The weight-length relation was also estimated, using the mean weights by age adopted by the WG for the long–term projections (ICES, 1998a), as being:

$$W(g) = 0.011\ L(cm)^{2.90}$$

The mortality of this stock is characterized as :

- A constant natural mortality coefficient during all the exploitable phase : $M = 0.15$ year^{-1}.
- Fishing mortality coefficients in 1996 (ICES, 1998a) are different with age as :

Age (year)	**1**	**2**	**3**	**4**	**5**	**6**	**7**	**8**	**9**	**10**
F_i (year^{-1})	0.24	0.26	0.10	0.08	0.06	0.09	0.12	0.14	0.18	0.24

1. Organize the calculations in a spreadsheet, in order to calculate, for each age of the life of the cohort (from age 1 to 10):

   a) The number at the beginning of the age

   b) The individual weight at the beginning of the age

   c) The biomass at the beginning of the age

   d) The number of deaths during the age

e) The mean number of survivors during the age

f) The mean individual weight during the age

g) The mean biomass during the age

h) The Catch in number during the age

i) The Catch in weight during the age

2. Determine:

   a) The cumulative number of survivors during all the exploitable phase

   b) The cumulative biomass of the cohort during all the exploitable life

   c) The total catch, in numbers, from the cohort during all the exploitable life;

   d) The total catch, in weight, from the cohort during all the exploitable life;

   e) The mean weight of the individuals caught, during all the exploitable life;

   f) The mean weight of the individuals of the cohort, during all the exploitable life.

GROUP II

Present the histograms of :

1. Mean numbers of survivors of the cohort in each age, during all the exploitable life.
2. Mean biomasses of the cohort at each age, during all the exploitable life.
3. Catches in number by age of the cohort, during all the exploitable life.
4. Catches in weight by age of the cohort, during all the exploitable life.

GROUP III

Suppose now, that one intends to analyse the case in which all the fishing mortality coefficients were 30 percent bigger than those indicated in the table of Group I.

1. Calculate the characteristics of the Group I-2 that could be obtained from the same cohort during all its life, in this alternative situation.
2. Compare the values of the characteristics obtained under these conditions with those obtained in Group I-2, calculating the percentage of variation of those characteristics in relation to the corresponding values of the previous situation.

8.9 COHORT DURING ITS LIFE – SIMPLIFICATION OF BEVERTON AND HOLT MODEL (3.7)

GROUP I

The recruitment to the exploitable phase of horse-mackerel, *Trachurus trachurus*, in Ibero-Atlantic waters (Div. VIIIc and IXa) occurs at age 1.

The recruitment at the exploitable phase was simplified by adopting the age t_c=2 year.

The parameters of the von Bertalanffy equation estimated for this stock are the following :

$L_\infty = 34.46$ cm
$K = 0.225$ year^{-1}
$t_o = -1.66$ year

The weight-length relation: $W(g) = 0.011\ L(cm)^{2.90}$

The mortality of this stock is characterized as :

- A constant natural mortality coefficient during all the exploitable phase $M = 0.15$ year^{-1};
- Fishing mortality coefficient, $F = 0.14$ year^{-1}, constant during all the exploited phase.

1. Calculate, using the simplification of Beverton and Holt :

    a) The recruitment R_c to the exploited phase.

    b) The number of deaths during all the exploited life.

    c) The cumulative number of survivors during all the exploited life.

    d) The cumulative biomass during all the exploited life.

    e) The catch in number during all the exploited life.

    f) The catch in weight during all the exploited life.

    g) The mean weight of the individuals of the cohort during all the exploited life.

    h) The mean weight of the individuals caught during all the exploited life.

GROUP II

The data presented in Section 8.8 – Group I show a great variety in the values of F. However, in Section 8.9, the simplification of an F constant was adopted, $F = 0.14$ year^{-1}. The purpose, now, is to compare the results from Section 8.8 with those obtained in this exercise, using the simplification of Beverton and Holt.

1.

a) Determinate the cumulative number, the cumulative biomass, the total catch in number, the total catch in weight and the mean weight in the catch.

b) Compare the results with those obtained in Section 8.8 (Group I-2).

2.

a) Using a 30 percent bigger value of F calculate the cumulative number, the cumulative biomass, the total catch in number, the total catch in weight and the mean weight in the catch and compare with the corresponding values obtained in the Group I-1.a).

b) Compare the percentages of variation of the characteristics obtained in 2.a) with those obtained in Section 8.8 (Group III).

8.10 STOCK – SHORT–TERM PROJECTION (4.3)

The mortality parameters and the exploitation pattern of the Iberic Stock of hake, *Merluccius merluccius*, from Divisions VIIIc and IXa of ICES were estimated by the Assessment Working Group of ICES (ICES, 1998b) as :

- Natural Mortality Coefficient = 0.20 $year^{-1}$
- Fishing Mortality Level in 1996 = 0.24 $year^{-1}$

Exploitation pattern in 1996

age	**0**	**1**	**2**	**3**	**4**	**5**	**6**	**7**	**8**
s_I	0.00	0.09	0.29	1.31	1.25	1.12	1.32	1.55	1.55

The growth parameters were estimated as being :

Growth parameters of von Bertalanffy (CE, 1994):	**Weight length relation, W_t(g)= a.L(cm)^b (Cardador, 1988):**
L_∞ = 100 cm K = 0.08 $year^{-1}$ t_0 = -1.4 year	a = 0.004 b = 3.2

At the beginning of 1996 the considered stock had the following age structure, with i representing the age and N_i the number of survivors at the beginning of the age i, expressed in millions of individuals :

age	**0**	**1**	**2**	**3**	**4**	**5**	**6**	**7**	**8**
N_i	83	27	41	30	22	11	6	3	5

GROUP I

1. Estimate the individual weights at the beginning of each age, the total number of individuals and the total biomass of the stock at the beginning of the year.
2. Estimate the individual mean weight, the mean number of survivors, the mean biomass, the total catch in number, the total catch in weight and the mean weight in the catch and in the stock, during 1996.
3. Supposing there was a recruitment in 1997 equal to 100 million individuals, calculate the following, for the beginning of 1997:

   a) The age structure of the stock, in number

   b) The age structure of the stock, in biomass

   c) The total number of individuals of the stock

   d) The total biomass of the stock

(notice that the stock at the beginning of 1997 is equivalent to the stock at the end of 1996, except for the 1997 recruitment).

GROUP II

The scientists drew attention to the fact that the fishing level in 1996 was very high and should, therefore, be reduced. They suggested a reduction of about 40 percent so that one could get adequate catches in weight and biomasses in the future.

As an alternative to improve the exploitation of this stock, they also suggested increasing the mesh size of the fishing nets.

1. Fishing managers asked the scientists to evaluate the mean biomass of the stock, the catch in number, the catch in weight and the mean weight in the catch during 1997 in case of :

   a) Maintaining the 1996 fishing pattern (status quo situation)

   b) Reducing the 1996 fishing level by 40 percent.

   Making the necessary calculations in order to estimate a) and b), present the results and comment on the changes in the catch in weight and mean biomass, relative to 1996.

2. Fishing managers also asked the scientists to evaluate the catch in weight and the mean biomass that would result from maintaining the fishing level of 1996 in 1997, but using a new mesh size of the fishing net. To solve this question, the scientists proposed the following new exploitation pattern:

age	0	1	2	3	4	5	6	7	8
s_i	0.00	0.00	0.03	1.50	1.65	1.75	1.80	1.80	1.80

Make the calculations, present the results and make your comments on the changes in the catch in weight and the mean biomass relative to 1996.

3. Evaluate the effects on the catch in weight and on the mean biomass, resulting from the simultaneous adoption (in 1997) of the reduction by 40 percent of the fishing level in 1996 and the introduction of the new exploitation pattern. Present the results and make your comments .

8.11 STOCK – LONG–TERM PROJECTION (4.4)

The mortality parameters of the Iberic Stock (Div. VIIIc e IXa) of Sardine, *Sardina pilchardus,* were estimated by the Assessment Working Group of ICES (ICES, 1997b) as :

- Natural Mortality Coefficient = 0.33 $year^{-1}$
- Fishing Mortality Level in 1996 = 0.56 $year^{-1}$

Assume that the exploitable phase occurs from the beginning of age zero until the end of age six.

Relative pattern of exploitation, s_i, during 1996

s_0	s_1	s_2	s_3	s_4	s_5	s_6
0.21	0.41	0.79	1.18	1.34	1.43	1.68

The parameters of the individual growth and the weight length relation of this stock were estimated (Pestana, 1989) as :

Von Bertalanffy growth parameters	**Weight length relation** $W = a.L^b$ **with** L **in cm and** W **in g**
$L_\infty = 22.3$ cm	a = 0.0044
$K = 0.40$ year^{-1}	b = 3.185
$t_o = -1.6$ year	

GROUP I

1. Calculate the evolution in number and the biomass of a cohort during its life at the beginning of each age, supposing that growth, natural and fishing mortality parameters are the given values.
2. The recruitment at age 0 in 1996 was estimated as being 4300 million individuals. Calculate the cumulative number, the cumulative biomass, the catch in number and in weight, during all the life of the cohort.

GROUP II

It was estimated that, at the beginning of 1996, the considered stock had the following age structure, in number, representing i the age and N_i the number of survivors at the beginning of age i, expressed in millions of individuals :

I	**0**	**1**	**2**	**3**	**4**	**5**	**6**
N_I	4300	279	591	233	561	384	180

To make a long–term projection of the stock, all the mortality and growth parameters will be considered stable during the following years. Consider, also, that the recruitment in those future years will be equal to the recruitment of 1996.

1. Based on these assumptions, project the numbers of survivors at the beginning of each year and age, until 2006.
2. Compare the structures of the stock in the years 2003 and 2006.
3. Compare the evolution of the 1996 cohort with the structure of the stock at the beginning of 2003.

8.12 STOCK – RECRUITMENT RELATION (4.5)

During the meeting of the assessment Working Group (ICES, 1998b) of the Iberic stock (Div. VIIIc and IXa) of hake, *Merluccius merluccius,* the following stock parameters were estimated for the period 1982-1996:

Year	N (age 0) (million)	Spawning Biomass (thousand tons)
1982	125	59.8
1983	107	61.4
1984	136	58.8
1985	97	44.1
1986	104	26.4
1987	97	24.2
1988	84	22.8
1989	56	18.9
1990	59	19.4
1991	69	20.5
1992	86	21.5
1993	70	21.0
1994	63	16.5
1995	32	15.2
1996	83	18.0

The annual recruitment at the exploitable phase is considered as being the number of individuals with age 0.

1. Draw the dispersion graph of the resulting recruitments, against the parental spawning biomass.

2. The parameters of Shepherd, Ricker, Beverton & Holt and Deriso S-R relations were estimated and are shown in the following table, as well as the respective determination coefficients, r^2:

Parameters	Shepherd	Ricker	Beverton & Holt	Deriso
α (R/g)	3.50	4.43	4.91	4.40
k (1000 tons)	64.94	78.13	45.39	106.27
C	3.52			0.896
r^2	0.71	0.68	0.66	0.75

a) Calculate the expected recruitments in each year of the period 1982-1996 using the four S-R models.

b) The determination coefficient, r^2, can be used as an indicator of a good or bad adjustment of the model to the observed data, depending on the number of observations (r^2 can be interpreted as a percentage of variation of the observed points that is explained by the model. Values close to 1 indicate a good adjustment and values close to zero indicate a bad adjustment). Using this indicator, make your comments about the adjustments of each model.

8.13 F_{max} (5.2.1)

Consider the stock of cod fish, *Gadus morhua,* in the Irish sea (Div. VIIa). The following mortality and biological parameters were estimated by the Working Group of ICES (ICES, 1998c):

Natural Mortality Coefficient: $M = 0.20$ year^{-1}

Fishing Level in 1996: $F_{96} = 0.58$ year^{-1}

Mean weight (kg) in the catch and in the stock :

Age	**0**	**1**	**2**	**3**	**4**	**5**	**6**	**7**
$\overline{W}_i$	0.001	0.883	1.778	3.597	5.695	7.904	8.502	9.200

GROUP I

Assume that the stock was constituted by the age groups 0 to 7 years and that it was exploited with the following exploitation pattern :

Age	**0**	**1**	**2**	**3**	**4**	**5**	**6**	**7**
s_i	0.80	0.90	0.96	1.00	1.00	1.00	1.00	1.00

1. Calculate, for 1000 recruits, the long–term annual catch in weight and the annual mean biomass, corresponding to the fishing level of 1996.
2. Adopting the factor F_{factor} between 0 and 2.5 year^{-1} with intervals of 0.1 year^{-1} draw the curve of the annual catch in weight against F_{factor}. Represent, in the same graph, the curve of the mean biomass against F_{factor}.
3. Calculate the biological reference point F_{max}.

GROUP II

1. Answer the previous questions, but considering the following exploitation pattern.

Relative pattern of exploitation adopted by the Working Group:

Age	0	1	2	3	4	5	6	7
s_i	0.00	0.20	0.96	1.30	1.12	0.67	0.58	0.58

a) Compare the value of F_{max} of this exercise with the one obtained in the exercise in Group I.

b) Calculate the biological reference point F_{max} knowing that the Working Group considers the group of age 7 as a group of cumulative ages (7+). The mean weights in the catch and in the stock presented at the beginning of the text are maintained, except for the last group of ages which will now be the age group 7+ with the mean weight equal to 10.873 kg. The exploitation pattern is also maintained with the value 0.58 for the group 7+.

c) Compare the two calculated values of F_{max}, considering the last age group as 7 or as 7+.

8.14 $F_{0.1}$ (5.2.2)

Consider the Iberic stock (Div. VIIIc e IXa) of four-spot megrim, *Lepidorhombus boscii*.

The following parameters were estimated by the Working Group of ICES (ICES, 1998b):

Natural Mortality Coefficient: $M = 0.20$ year^{-1}

Fishing mortality in 1996: $F_{96} = 0.36$ year^{-1}

Exploitation pattern:

Age	1	2	3	4	5	6	7+
s_i	0.06	0.43	0.89	1.65	1.66	1.22	1.22

Mean weight (kg) in the catch and in the stock:

Age	1	2	3	4	5	6	7+
$\overline{W}_i$	0.037	0.067	0.086	0.109	0.144	0.188	0.244

1. Adopting the value 1000 for the recruitment to the fishing area, calculate the long–term annual catch in weight, the annual mean biomass, and the annual mean weight in the catch, corresponding to the fishing mortality level of 1996.
2. Adopting the factor F_{factor} between 0 and 2.5 year^{-1} with intervals of 0.1 year^{-1} draw the curve of the annual catch in weight against F_{factor}. Represent, in the same graph, the curve of the mean biomass against F_{factor}.
3. Calculate the biological reference point $F_{0.1}$
4. Calculate the biological reference point F_{max}
5. In the graph of the curves of the annual catch in weight and of the mean biomass against F_{factor}, mark the Biological reference points $F_{0.1}$ and F_{max} which were previously calculated. Make your comments.
6. Calculate the long–term mean biomass, the catch in weight and the mean weight in the catch for $F_{0.1}$. Compare those characteristics with the values obtained in question 1 and make your comments.

8.15 F_{med} and F_{MSY} (5.2.3) and (5.2.4)

During the meeting of the Assessment Working Group (ICES, 1998b) on the Iberic stock (Div. VIIIc and IXa) of hake, *Merluccius merluccius,* the following population parameters were estimated :

Natural mortality coefficient: $M = 0.2$ year^{-1};

Fishing mortality in 1996: $F_{96} = 0.24$ year^{-1}

Mean weight in the catch (g):

Age (year)	**0**	**1**	**2**	**3**	**4**	**5**	**6**	**7**	**8+**
$\overline{W}_i$(g)	4	37	106	205	358	517	706	935	1508

Maturity ogive (percent):

Age (year)	**0**	**1**	**2**	**3**	**4**	**5**	**6**	**7**	**8+**
% mature$_i$	0	0	1	6	20	49	76	91	100

For the long–term projections, the Working Group adopted the mean exploitation pattern for the period 1994-1996, as shown in the next table :

Age (year)	**0**	**1**	**2**	**3**	**4**	**5**	**6**	**7**	**8+**
s_i	0.001	0.11	0.398	1.3	1.261	1.019	1.473	1.874	1.874

1. Calculate the long–term annual mean biomass, the spawning biomass, the annual catch in weight and the mean weight in the catch.

2. Draw the curve of the annual catch in weight and the mean biomass against F, for values of F_{factor} between 0 and 2.5 $year^{-1}$, with intervals of 0.1 $year^{-1}$. Calculate the TRPs F_{max} and $F_{0.1}$.

3. Calculate the biological reference point F_{med} , knowing the recruitments (million individuals at age 0) and the spawning biomasses (thousand tons) between 1982 and 1996, estimated by the Working Group, presented in the following table :

Year	**N (age 0) (million)**	**Spawning biomass (1000 tons)**
1982	125	59.8
1983	107	61.4
1984	136	58.8
1985	97	44.1
1986	104	26.4
1987	97	24.2
1988	84	22.8
1989	56	18.9
1990	59	19.4
1991	69	20.5
1992	86	21.5
1993	70	21.0
1994	63	16.5
1995	32	15.2
1996	83	18.0

4. Calculate the long–term mean biomass, the spawning biomass, the catch in weight and the mean weight in the catch for F_{med}. Compare with the values obtained in question 1 and make your comments.

5. Adopting the Ricker stock-recruitment relation, estimated by the Working Group (α = 4.43 R/ kg and K = 78.13 thousand tons) , calculate F_{MSY}, B_{MSY} and Y_{MSY}. and compare the different F-target estimated.

8.16 MBAL AND B_{loss} (5.3.4 & 5.3.5)

During the meeting of the Assessment Working Group (ICES, 1998b) on the Iberic stock (Div. VIIIc and IXa) of hake, *Merluccius merluccius,* the recruitment (million individuals at age 0) and the spawning biomass (thousand tons) was estimated for the period 1982-1996. The values obtained are presented in the following table :

Year	N (age 0) (million)	Spawning biomass (1000 tons)
1982	125	59.8
1983	107	61.4
1984	136	58.8
1985	97	44.1
1986	104	26.4
1987	97	24.2
1988	84	22.8
1989	56	18.9
1990	59	19.4
1991	69	20.5
1992	86	21.5
1993	70	21.0
1994	63	16.5
1995	32	15.2
1996	83	18.0

1. Using the spawning biomasses and the resulting recruitments, calculate the Biological Reference Limit Points, MBAL and B_{loss}.

8.17 F_{loss} and F_{crash} (5.3.5 & 5.3.6)

GROUP I

During the meeting of the Assessment Working Group (ICES, 1998b) on the Iberic stock (Div. VIIIc and IXa) of hake, *Merluccius merluccius*, the recruitment (million individuals at age 0) and the spawning biomass (thousand tons) was estimated for the period 1982-1996. The values obtained are shown in the following table :

Year	N (age 0) (million)	Spawning biomass (thousand tons)
1982	125	59.8
1983	107	61.4
1984	136	58.8
1985	97	44.1
1986	104	26.4
1987	97	24.2
1988	84	22.8
1989	56	18.9
1990	59	19.4
1991	69	20.5
1992	86	21.5
1993	70	21.0
1994	63	16.5
1995	32	15.2
1996	83	18.0

The Shepherd S-R relation was adjusted to the pairs of values in the table (r^2=0.71), and the relation parameters are the following :

$\alpha = 3.5$ Kg^{-1}
$k = 64.94$ thousand tons
$c = 3.52$

1. Draw the dispersion graph of the resulting recruitments, against the parental spawning biomasses.
2. Calculate the expected recruitments in each year of the period 1982-1996 according to the Shepherd S-R model and in the previous dispersion graph, draw the respective curve.

GROUP II

1. Calculate the annual catch in weight and the spawning biomass per recruit for the stock of hake, using the mortality and biological parameters estimated by the Working Group for the long–term projections (given in Section 8.15), namely :

Natural mortality coefficient : $M = 0.2\ year^{-1}$
Fishing mortality in 1996: $F_{96} = 0.24\ year^{-1}$

Mean weight in the catch (g):

Age (year)	**0**	**1**	**2**	**3**	**4**	**5**	**6**	**7**	**8+**
$\overline{W}_i$	4	37	106	205	358	517	706	935	1508

Maturity ogive (percent):

Age (year)	**0**	**1**	**2**	**3**	**4**	**5**	**6**	**7**	**8+**
% mature$_i$	0	0	1	6	20	49	76	91	100

Mean relative pattern of exploitation of the period 1994-1996:

Age (year)	**0**	**1**	**2**	**3**	**4**	**5**	**6**	**7**	**8+**
s_i	0.001	0.11	0.398	1.3	1.261	1.019	1.473	1.874	1.874

GROUP III

1. Using the results of Groups I and II, calculate biological reference Limit-Points, F_{loss} and F_{crash}.

8.18 PRODUCTION MODELS (EQUILIBRIUM) – SCHAEFER (6.7.1)

The following table presents the annual total catch Y, (t) and mean biomass $\bar{B}$, (t) for the fishery of the Iberic stock (Div. VIIIc and IXa) of sardine, *Sardina pilchardus,* between 1977 and1996, used by the Working Group of ICES (ICES, 1998a).

Year	Y (t)	$\bar{B}$ (t)
1977	125750	750289
1978	139990	759192
1979	153441	763313
1980	191682	804765
1981	214133	842091
1982	204504	802573
1983	181139	713376
1984	202686	794856
1985	204107	810539
1986	180606	679808
1987	168825	547179
1988	158540	481295
1989	137126	431719
1990	139157	368099
1991	127756	316365
1992	126054	453161
1993	138795	539096
1994	132800	416842
1995	121384	368158
1996	111431	246037

GROUP I

1. Calculate F_i corresponding to each year i.
2. Calculate the biomasses, B_i, at the beginning of each year. (Use the procedure proposed by Schaefer, that is, the biomass at the beginning of a year is approximated by the arithmetic mean of the mean biomasses of the previous and the following year).
3. Calculate the equilibrium catches, Y_E, which would correspond to the observed values of F.

4. Calculate the equilibrium mean biomasses, $\overline{B}_E$, which would correspond to the observed values of F.

5. Draw the graph of $\overline{B}_E$ against F_i.

GROUP II

The Schaefer model was adjusted and the following parameters were estimated :

$k = 1562851$ t
$r = 0.426$ year^{-1}

1. Calculate the equilibrium biomasses and the equilibrium catches corresponding to the fishing levels observed in each year, using the Schaefer model.

2. Draw, in the graph of Group I point 5, the equilibrium biomasses calculated with the Schaefer model.

3. Calculate F_{MSY}, B_{MSY} and Y_{MSY}

4. Calculate $F_{0.1}$, $B_{0.1}$ and $Y_{0.1}$.

8.19 PRODUCTION MODELS (EQUILIBRIUM) (6.7.1 & 6.7.2)

The following table presents the annual total catches and the corresponding total fishing efforts of a shrimp stock in the Arabian Sea during the period 1969 to 1978 (Sparre and Venema, 1992).

Year	Catch, Y (t)	Total effort (1000 days)
1969	546.7	1.224
1970	812.4	2.202
1971	2493.3	6.684
1972	4358.6	12.418
1973	6891.5	16.019
1974	6532.0	21.552
1975	4737.1	24.570
1976	5567.4	29.441
1977	5687.7	28.575
1978	5984.0	30.172

GROUP I

1. Draw a graph of the annual abundance index against the corresponding fishing mortality index.

2. The Fox model was adjusted to the data. The following parameters were obtained :

$a = 6.150$ $b = -0.028$ with a determination coefficient, $r^2 = 0.78$.

a) Calculate and draw the curves of the equilibrium conditions of the abundance index and of the total catch against fishing effort.

b) Determine the target-points MSY and 0.1.

c) Determine the parameters (kq), (r/q) and (kr).

GROUP II

Knowing that the adjustment of the Schaefer model to the same set of data, resulted in the following values of the parameters :

$a = 444.454$
$b = - 8.055$
$(r^2 = 0.77)$

1. Repeat the calculations of the previous Group.

8.20 PRODUCTION MODELS – SHORT–TERM PROJECTION (6.8)

The following table presents the annual total catches (in tons) and the corresponding catches by fishing unit effort (kg/ fishing day of the fleet PESCRUL) of the stock of Deepwater rose shrimp, *Parapenaeus longirostris,* of the Algarve during the period 1983 to 1994 (Mattos Silva, 1995).

Year	Y (t)	cpue (kg/day)
1983	538	235
1984	638	131
1985	431	63
1986	99	22
1987	37	8
1988	62	21
1989	437	77
1990	146	28
1991	126	26
1992	53	25
1993	91	41
1994	232	66

Using the following parameters of the integrated Fox model and Yoshimoto & Clarke (1993): k= 1580 t, q=0.39 thousand $days^{-1}$ and r=0.55 $year^{-1}$:

1. Project the cpue and the catch in weight for the year 1995, supposing that the fishing effort will be maintained equal to that of 1994 (situation status quo).
2. Determine the target reference points, Y_{MSY}, $\overline{B}_{MSY}$ and F_{MSY} and the indices $\overline{U}_{MSY}$ and f_{MSY}.
3. Determine the target reference points, $Y_{0.1}$, $\overline{B}_{0.1}$ and $F_{0.1}$ and the indices $\overline{U}_{0.1}$ and $f_{0.1}$.
4. Determine the percentages of the carrying capacity corresponding to the target points F_{MSY} and $F_{0.1}$
5. Based on the results obtained in the previous questions, make your comments on the state of the stock and of its exploitation.
6. Suppose that one intends to reduce the fishing effort in 1995 by about 20 percent relative to the effort of 1994. Project the catch in weight for 1995 and present the variations resulting from that effort reduction on the catch in weight and on the biomass.

8.21 SIMPLE LINEAR REGRESSION – ESTIMATION OF THE PARAMETERS OF THE W-L RELATION AND GROWTH PARAMETERS (FORD-WALFORD, GULLAND AND HOLT AND STAMATOPOULOS AND CADDY) (7.2)

GROUP I

Consider the following 10 pairs of values of x and y:

x_i	2	6	7	8	11	15	16	18	19	21
y_i	13	40	52	56	78	105	111	130	132	149

1. Estimate the constants A and B of the straight line.
2. Estimate the values of Y corresponding to the given values of x.
3. Calculate the determination coefficient r^2.
4. Draw a graph with the observed values and with the estimated line. Observe the adjustment and say if you consider the linear model adequate.

GROUP II

The data presented on the following table represent the individual weights by length class of European anglerfish, *Lophius budegassa* samples from the Iberic coast in 1994.

L_i (cm)	$\overline{W}_t$ (g)	L_i (cm)	$\overline{W}_t$ (g)
20-	129	50-	1685
22-	163	52-	1896
24-	219	54-	2107
26-	265	56-	2345
28-	320	58-	2569
30-	397	60-	2848
32-	486	62-	3126
34-	545	64-	3407
36-	664	66-	3700
38-	773	68-	4056
40-	890	70-	4411
42-	1027	72-	4764
44-	1122	74-	5203
46-	1334	76-	5587
48-	1503	78-	5982

1. Using the simple linear regression model, estimate the parameters of the weight-length relation for this stock, considering that $W_{central}=W_{mean}$. (Notice that the ln $W_{central}$ is linear to the ln $L_{central}$).

GROUP III

The data presented in the following table represent the mean length (cm) of the individuals at the beginning of the age (years), obtained from direct age reading of the individuals of the stock of European anglerfish, *Lophius budegassa,* (ICES Div. VIIIc and IXa), Section 8.7

T	L_t	t	L_t
1	9.2	7	44.4
2	16.5	8	49.0
3	22.9	9	52.3
4	28.8	10	55.0
5	34.7	11	60.8
6	38.6	12	63.4

From this data and using the simple linear regression model, estimate the growth parameters K and L_∞ with the expressions :

1. Ford-Walford (1933-1946)
2. Gulland and Holt (1959)
3. Stamatopoulos and Caddy (1989). (The expressions for 1), 2) and 3), were studied in the Chapter concerning Individual Growth).
4. Comment on the results obtained in the previous questions with the values of the parameters given in Section 8.7.

8.22 MULTIPLE LINEAR MODEL – REVISION OF MATRICES – (7.3) ESTIMATION OF THE PARAMETERS OF FOX INTEGRATED MODEL (IFOX)

REVISION OF MATRICES

GROUP I

Consider the matrices A and B :

$$A = \begin{matrix} 2\,3\,0\,1 \\ 1\,1\,4\,1 \\ 0\,4\,2\,2 \\ 1\,0\,3\,3 \end{matrix} \qquad B = \begin{matrix} 1\,1\,0\,3 \\ 1\,3\,2\,5 \\ 2\,1\,6\,0 \\ 2\,2\,1\,0 \end{matrix}$$

1. Using a spreadsheet, calculate: A + B, A * B, Det(A), Det(B), A^{-1} e B^{-1}
2. Show that $(A.B)^{-1} = B^{-1}.A^{-1}$
3. Show that $(A.B)^T = B^T.A^T$

GROUP II

Let the Matrices:

$$M_{(4,4)} = (1/4) \begin{matrix} 1\,1\,1\,1 \\ 1\,1\,1\,1 \\ 1\,1\,1\,1 \\ 1\,1\,1\,1 \end{matrix} \qquad O_{(4,4)} = \begin{matrix} 0\,0\,0\,0 \\ 0\,0\,0\,0 \\ 0\,0\,0\,0 \\ 0\,0\,0\,0 \end{matrix} \qquad I_{(4,4)} = \begin{matrix} 1\,0\,0\,0 \\ 0\,1\,0\,0 \\ 0\,0\,1\,0 \\ 0\,0\,0\,1 \end{matrix}$$

1. Verify that the matrix null 0 is idempotent.
2. Verify that the matrix identity I is idempotent.
3. Verify that the matrix M is idempotent.

4. What are the traces of M and I?

5. Calculate the ranks, r , of M and of I.

6. What is the value of the determinant of M and I?

GROUP III

1. Verify that the product Mx, where x is the vector given by $x^T = (3\ 4\ 8\ 1)$, is a vector with all elements equal to the arithmetic mean, $\bar{x}$, of the 4 elements of vector x.

2. Verify that (I-M)x is the vector of the deviation.

3. Verify that the sum of the squares of x_i, $\Sigma(x_i^2)$ can be written as : $x^T . x$

4. Verify that the sum of the squares of the deviations, $\Sigma(x_i - \bar{x})^2$, can also be written in a matricial form, as : $x^T(I-M)x$

GROUP IV

1. Consider the vector $x = \begin{matrix} 2+\theta \\ 3\theta \\ 5-\theta \end{matrix}$ where θ is an unknown parameter.

a) Write the derivative $\frac{dx}{d\theta}$ of the vector x

b) Calculate $x^T x$

c) Calculate $\frac{d}{d\theta}(x^T x)$

d) Show that $\frac{d}{d\theta}(x^T x) = 2\,(\frac{dx}{d\theta})^T x$

2. Consider the vector $x = \begin{matrix} 2+\ 4\theta_1 -\ 5\theta_2 \\ 1+\ \theta_1 +\ \theta_2 \\ \theta_1^2 + 4\theta_2 \end{matrix}$ where θ_1 and θ_2 are two unknown constants.

a) Write the derivative matrix $\frac{\partial x}{\partial \theta}$ (take θ_1 and θ_2 as variables)

b) Calculate $x^T x$

c) Transpose $\frac{\partial}{\partial \theta}(x^T x)$

d) Show that the transposed matrix $\frac{\partial}{\partial\theta}(x^T x) = 2\,(\frac{\partial x}{\partial\theta})^T x$

GROUP V

Consider the following system of 2 equations with 2 unknowns;

$$5 = 2A + 3B$$
$$4 = A - 2B$$

1. Show that the equation system can be written in matrix form as,

$$Y_{(2,1)} = X_{(2,2)}\,\theta_{(2,1)}$$

where Y is the vector of the independent terms (5 e 4) of the system,

θ is the vector of the unknowns A and B
and X is the matrix of the coefficients of the unknowns

2. Verify that the solution of the system can be given as $\theta = (X^T X)^{-1} X^T Y$
3. Show that X is a square, non singular matrix, and then that the solution of the system can be $\theta = X^{-1}Y$

ESTIMATION OF THE PARAMETERS OF THE YOSHIMOTO AND CLARKE MODEL (1993)

4. Estimate the parameters k, q and r, of the Fox integrated model (IFOX) and of Yoshimoto & Clarke (1993) using the following data :

Year	Y (t)	CPUE (kg/day)
1983	538	235
1984	638	131
1985	431	63
1986	99	22
1987	37	8
1988	62	21
1989	437	77
1990	146	28
1991	126	26
1992	53	25
1993	91	41
1994	232	66

which represent the total annual catches (in tons) and the respective catches by fishing effort unit (kg/fishing day of the fleet PESCRUL) of the stock of Deepwater rose shrimp, *Parapenaeus longirostris* of the Algarve during the period 1983 to 1994 (Mattos Silva, 1995).

Comment on the obtained results comparing them with those presented in Section 8.20.

8.23 NON LINEAR REGRESSION – ESTIMATION OF THE GROWTH PARAMETERS AND OF THE S– R RELATION (GAUSS–NEWTON METHOD) (7.4)

The data in the following table represent the mean length (cm) of the individuals at the beginning of the age (years), obtained from direct age reading of individuals of the stock of anglerfish, *Lophius budegassa,* (Div. VIIIc andIXa), Section 8.7.

t	L_t	t	L_t
1	9.2	7	44.4
2	16.5	8	49.0
3	22.9	9	52.3
4	28.8	10	55.0
5	34.7	11	60.8
6	38.6	12	63.4

GROUP I

1. Represent graphically the values of L_t against t.
2. Estimate the growth parameters K, L_∞ and t_0 for the Bertalanffy growth model using a non-linear regression model.
3. Estimate the values of L_t corresponding to given values of t and mark on a graph the observed values and the estimate curve. Comment on the adjustment.
4. Compare the results obtained in the previous questions with the values of the parameters given in Section 8.7.

GROUP II

Using the data on spawning biomass and on recruitments of hake, presented in Section 8.12, estimate the parameters of the Beverton and Holt, Ricker, Deriso and Shepherd S-R models.

1. Compare the obtained values with those presented in Section 8.12.

8.24 ESTIMATION OF M (7.6)

GROUP I

1. Estimate the 5 percent Tanaka survival curve for natural mortality coefficients between 0.0 and 5.0 year^{-1} and longevities between 0 and 30 years.
2. Calculate the values of M corresponding to the longevity 1, 2, 3, 10, 15 and 30 years.

GROUP II

The parameters of von Bertalanffy equation for the Iberic stock (Div. VIIIc and IXa of ICES) of horse-mackerel, *Trachurus trachurus,* are the following :

$L_\infty = 34.46$ cm, TL (carapace length)
$K = 0.225$ year^{-1}

1. Estimate the value of M for this stock, knowing that from 1985-95 in the distribution area of this stock in the Iberic Peninsula, the mean temperature of the sea water at the surface, was : $\overline{T} = 13^\circ C$

GROUP III

Consider a certain fishing stock for which the first maturity mean age was estimated as 2.3 year.

1. Obtain an approximate estimate of M for this stock.

GROUP IV

The reproductive biology of the stock of Atlantic mackerel, *Scomber scombrus* was studied and it was estimated that the mean gonadsomatic index (gonad weight/total weight) of the mature females in the spawning period was 0.13.

1. Estimate an approximate value for the natural mortality coefficient M for this stock, assuming that it is constant for all the ages and years.

GROUP V

The scientists responsible for the evaluation of the stock of a certain fishing resource, implemented an acoustic cruise every year in January, to estimate the abundance of the stock by age classes. Fishery statistics are also needed to estimate the catch by age during the year.

The following table presents the estimations of the abundance of the stock, by age classes, obtained on the cruises in 1993 and 1994, as well as the structure of the catches during 1993.

Age	Number of survivals, N, in million		Total catch in 1993 (million)
	January 1993	January 1994	
2	243	353	11
3	99	189	15
4	86	67	20
5	37	52	9
6	13	22	3
7	6	8	1

Although this information is available, the scientists responsible for this stock have difficulties applying the assessment models to the resource because they do not have an estimation of M.

1. So, estimate the natural mortality coefficient for this resource in 1993, and help the group of scientists responsible for the resource.

GROUP VI

The following table presents the data on fishing effort, in million trawl hours, and the total mortality coefficent, Z, for a certain fishery for the period 1987 to 1995.

Years	Effort (10^6 hours)	Z ($year^{-1}$)
1987	2.08	1.97
1988	2.80	2.05
1989	3.5	1.82
1990	3.6	2.32
1991	3.8	2.58
1992	-	-
1993	-	-
1994	9.94	3.74
1995	6.06	3.74

1. Determine M (natural mortality coefficient), assumed to be constant for the period 1987-1995.
2. Determine the catchability coefficient q.

8.25 ESTIMATION OF Z (7.7)

GROUP I

A swept area cruise allowed the scientists of the Marine Research Institute in Bergen, Norway, to estimate the abundance of the different age classes of the stock of cod fish, *Gadus morhua,* in January of 1995 (following table).

Age (years)	1	2	3	4	5	6	7	8	9	10	11	12	13	14	15
N_{95} (10^9)	1984	440	160	103	82	65	54	43	33	27	26	21	17	13	10

1. Represent on a graph, the logarithms of the numbers of survivors against the age.
2. Select the age interval from which the total mortality coefficient, Z, can be taken as constant.
3. Estimate the total mortality coefficient, Z, of the stock in January 1995.

GROUP II

The following table presents the mean catches by age, in number, of plaice, *Pleuronectes platessa,* per 100 trawl hours in two periods, 1929-1938 and 1950-1958.

Age (years)		2	3	4	5	6	7	8	9	10
C/f	1929-38	125	1355	2352	1761	786	339	159	70	28
C/f	1950-58	98	959	1919	1670	951	548	316	180	105

1. Estimate the total mortality coefficient, Z, of the stock in each of the periods.
2. Consider that the mean fishing effort on the North Sea plaice during the two periods was 5 million hours of trawl in 1929-1938 and 3.1 million hours of trawl in 1950-1958. Estimate for each period:

a) the natural mortality coefficient, M;

b) the catchability coefficient, q;

c) and the fishing mortality coefficient, F.

GROUP III

The following table presents the annual composition of the catches by age from 1988 to 1994, in millions of individuals, for a certain resource :

CATCHES (million individuals)

Age	1988	1989	1990	1991	1992	1993	1994
0	599	239	424	664	685	478	330
1	678	860	431	1004	418	607	288
2	1097	390	1071	532	335	464	323
3	275	298	159	269	203	211	243
4	40	54	75	32	69	86	80
5	6	9	13	18	8	25	31
6	1	8	3	5	5	3	8
7	6	0	1	0	1	1	1

1. Calculate the mean annual composition during 1988-1994.
2. Estimate Z, based on that mean composition.
3. Estimate Z, based on the mean age of the mean composition of the catch.
4. Estimate Z for each year of the given period.
5. Compare the annual Zs with the values of Z obtained in questions 2 and 3.

GROUP IV

The following table shows the length composition, in equilibrium, of a certain resource, with $L_\infty = 100$ cm and K = 0.2 year^{-1}.

Length class (cm)	**35-**	**40-**	**45-**	**50-**	**55-**	**60-**	**65-**	**70-**	**75-**	**80-**	**85-**	**90-**	**95-**
Catch (C_i) in million	7	10	20	51	46	44	41	36	33	28	23	17	8

1. Calculate the relative ages corresponding to the lower limit of each length class.
2. Determine the age interval corresponding to each length class.
3. From which class can one consider Z constant?

4. Determine Z using :

a) The catches in each class.

b) The cumulative catches.

c) The mean length in the catch.

5. Compare the values of Z obtained by the different methods of question 4.

GROUP V

The length compositions of the catches for three different periods of time are known for a certain fishing resource.

Period	Length classes (cm)	45-	50-	55-	60-	65-	70-	75-	80-	85-	90-	≥95
1960-69	Catch (C_i) in million	256	237	211	187	161	138	113	87	62	36	12
1970-79		268	226	180	141	105	76	50	30	15	6	1
1980-89		212	161	116	79	52	31	17	8	3	1	0

Consider the 45 cm length class as the first class completely recruited .
Adopt K = 0.3 year^{-1} and $L_\infty = 100\ cm$ as the von Bertalanffy growth parameters for this resource.

1. Estimate the values of the total mortality coefficient, Z, for each period and comment on the results.

8.26 AGE COHORT ANALYSIS (CA) (7.9.1)

GROUP I

1. Consider a stock and an interval of time i, (t_i, t_{i+1}). Knowing that for this interval of time:

M_i = 0.4 year^{-1}
T_i = 2.3 year
C_i = 230 million individuals

a) Adopt the value 0.5 year^{-1} for the fishing mortality coefficient for the interval and calculate the numbers of survivors at the beginning and end of the interval.

2. Consider the interval of time i, (t_i, t_{i+1}). Knowing that in this interval of time:

M_i = 0.6 year^{-1}
T_i = 0.9 year
C_i = 98 million individuals

Calculate the value of the fishing mortality coefficient, F_i, for the interval , taking the number of survivors, N_i, at the beginning of the interval, i, to be 172 million individuals.

3. Consider the interval of time (t_i, t_{I+1}). Knowing that in that interval of time :

$M_i = 0.5\ year^{-1}$
$T_i = 1$ year
$C_i = 42$ million individuals

a) Calculate the value of the fishing mortality coefficient for the interval, knowing that the number of survivors at the end of the year was $N_{i+1} = 85$ million individuals. Calculate the value of F_i using the Pope formula.

GROUP II

The data in the following table represent the catches in millions, of a cohort of hake, *Merluccius merluccius,* in the Iberic Peninsula waters.

Age (years)	**0**	**1**	**2**	**3**	**4**	**5**	**6**	**7**	**8**
C_i (million)	712	3941	8191	10311	5515	4149	3081	1185	549

Adopt a value of 0.2 $year^{-1}$ for the natural mortality coefficient, constant for all the ages.

1. Suppose that the value of the fishing mortality coefficient at the last age (8 years) was 1.0 $year^{-1}$. Calculate, by an iterative method and by the Pope method, for each age of the cohort :

a) The value of the fishing mortality coefficient.

b) The number of survivors at the beginning of the age.

c) Compare the results obtained by the two methods.

d) Represent, on a graph the values of F_i estimated against the age, and say what the recruitment of this cohort is at the exploited phase.

GROUP III

1. Aiming to analyse the influence of the chosen $F_{terminal}$, repeat the calculations of question 1 of Group II, using one of the previous methods, with 0.3 and 1.5 $year^{-1}$ for the value of $F_{terminal}$.

a) Draw a graph with the estimated values of F_i and N_i against the age.

b) Comment on the differences between the graphs for the different values of $F_{terminal}$.

2. Aiming to analyse the influence of the choice of M, repeat the calculations of question 1 of Group II, using one of the previous methods, for values of M of 0.1 and 0.4 $year^{-1}$.

a) Represent, on a graph, the estimated values of F_i and N_i against the age.

b) Comment on the differences between the graphs for the different values of M.

GROUP IV

The annual catches by age class, of a certain resource, for the years of 1985 to 1994, are presented in the following table.

Catches by age class (Million individuals)										
	Years									
Age (years)	**1985**	**1986**	**1987**	**1988**	**1989**	**1990**	**1991**	**1992**	**1993**	**1994**
0	67	88	104	290	132	90	63	38	52	90
1	532	1908	1841	1671	4172	1915	1284	906	541	704
2	2070	1756	4424	3178	2534	6320	2826	1911	1322	741
3	728	4016	2256	4042	2499	1972	4742	2115	1382	890
4	353	945	3309	1273	1926	1170	883	2102	896	540
5	97	439	733	1730	558	827	479	356	807	316
6	16	107	300	333	656	207	291	166	117	243
7	25	8	73	136	126	243	73	101	54	35
8	5	7	5	33	52	47	85	25	33	16

The *modus operandi* of the fishing fleet was constant during the period, but the number of vessels increased significantly. It is considered that, at present, the resource is intensively exploited.

Besides the information on the fishery, the estimates of the growth parameters of this resource and of the natural mortality coefficient are also available:

$L_\infty = 38.5$ cm
$K = 0.25$ year^{-1}
$t_o = -0.51$ year

a = 0.021 of the relation W(g)-L(cm)
b = 2.784 of the relation W(g)-L(cm)
$M = 0.3$ year^{-1}

1. Estimate the fishing mortality coefficient and the number of survivors at the beginning of the year for each age class and each year. Use the Pope Cohort Analyses method.

a) Start by selecting $F_{terminal} = 0.5$ year^{-1} for the last age of every year and for all the ages of the last year.

b) After analysing matrix F obtained in a), select new values for $F_{terminal}$ and repeat the application of Pope's method.

2. Besides the information given in the previous question, it is also known that the spawning takes place in a restricted period, around the beginning of the year. Research cruises using acoustic methods took place during the spawning period, in order to estimate the spawning biomass (kg/hour of trawl). The results obtained are shown in the following table :

Years	**1985**	**1986**	**1987**	**1988**	**1989**	**1990**	**1991**	**1992**	**1993**	**1994**
Spawning biomass Index	1270	1613	1629	1424	1300	1209	1000	718	476	326

The biological information collected during those cruises was also used to estimate the maturity ogive of the stock at the spawning period:

Age (years)	**0**	**1**	**2**	**3**	**4**	**5**	**6**	**7**	**8**
% Matures	0	1	20	50	80	100	100	100	100

a) Calculate the spawning biomass in the spawning period of each year from 1985 to 1994 using the results of the Cohort Analyses obtained in question 1.b.

b) Use the information of the acoustic cruises to tune the Cohort Analyses.

c) Comment on the tuning results.

8.27 LENGTH COHORT ANALYSIS (LCA) (7.9.2)

GROUP I

The following table presents the annual catch length composition, of a cohort of a resource with $L_\infty = 130$ cm and $K = 0.1$ year^{-1}.

Length classes(cm)	Catch, C_I (million)
6-	1823
12-	14463
18-	25227
24-	8134
30-	3889
36-	2959
42-	1871
48-	653
54-	322
60-	228
66-	181
72-	96
78-	16
84-	0

The natural mortality coefficient was estimated as being M = 0.3 $year^{-1}$.

1. Using the Pope method, and adopting E=0.5 as being the exploitation rate in the (78-) length class of the catch, estimate the number of survivors at the beginning of each length class, the fishing mortality coefficient F and the exploitation rate E in each class.
2. Calculate the mean number of survivors of the cohort.

GROUP II

The following tables 1 and 2 present the basic information on a hypothetical stock during the years 1985 to 1994.

1. Apply the slicing technique to the Catch matrix and comment on the validity of applying cohort analyses by ages.
2. Estimate the matrices [F] and [N] by length classes and years.
3. Calculate the matrix [Fsep] and comment on the hypothesis that the exploitation pattern can be considered to be constant during those years.

Table 1. Growth parameters of the von-Bertalanffy curve, L_∞ and K Natural Mortality Coefficient, M and constants a and b of the weight/length relation

Growth		Natural Mortality		Weight/length relation $w_i = a \cdot (L_i)^b$	
L_∞ (cm)	42	M ($year^{-1}$)	0.8	a	0.0023
K ($year^{-1}$)	0.5			b	3

Table 2. Catch matrix in thousands of individuals, by length classes and years in the period 1985-94

Age (sliced)	Length classes (cm)	Years									
		1985	1986	1987	1988	1989	1990	1991	1992	1993	1994
0	20-	35	41	30	17	49	69	34	61	46	29
	21-	338	400	292	167	472	662	327	593	442	276
	22-	805	952	699	400	1127	1575	777	1404	1053	657
	23-	1500	1766	1317	757	2108	2923	1436	2574	1962	1220
	24-	1901	2222	1702	985	2688	3678	1795	3175	2485	1535
	25-	2034	2357	1872	1093	2902	3900	1886	3276	2659	1627
	26-	1898	2175	1806	1067	2739	3600	1722	2925	2482	1502
1	27-	1951	1817	1228	1416	1445	2932	3376	1695	1785	2376
	28-	1664	1529	1091	1276	1250	2467	2801	1369	1523	1999
	29-	1382	1251	948	1125	1053	2018	2258	1071	1265	1636
	30-	1127	1003	812	980	873	1619	1782	818	1031	1312
	31-	900	787	684	841	710	1269	1372	607	823	1029
	32-	694	595	560	702	558	959	1017	432	635	778
2	33-	809	565	290	389	834	511	759	832	221	518
	34-	584	399	226	310	618	361	522	544	160	365
	35-	403	267	170	240	439	242	340	335	110	245
	36-	262	168	122	178	294	152	207	191	72	154
3	37-	165	168	66	71	175	214	93	128	75	46
	38-	86	84	40	45	96	107	44	55	39	23

Consider L_a= 20 cm and t_a=0
(Extracted from : Cadima, E. & Palma, C.,1997. Cohort Analysis from annual length catch compositions. Working document presented to the Working Group of the Demersal Stocks Assessment of the South Shelf, held in Copenhagen from 1-10 September, 1997.)

8.28 EXAMINATION - WRITTEN TEST

TRAINING COURSE ON FISH STOCK ASSESSMENT INSTITUTO DE INVESTIGAÇÃO DAS PESCAS E DO MAR (LISBON, 3 NOV. -10 DEC. 1997)

QUESTION 1

Consider a certain stock with the following parameters:

Natural Mortality Coefficient: $M = 0.20\ year^{-1}$

Fishing mortality in 1996: $F_{96} = 1.08\ year^{-1}$

Exploitation pattern :

Age	1	2	3	4	5	6	7+
s_I	0.07	0.23	0.33	0.49	0.97	1.00	1.00

Mean weight (kg) in the catch and in the stock:

Age	1	2	3	4	5	6	7+
$\overline{W}_I$	0.053	0.076	0.111	0.125	0.158	0.204	0.337

Maturity ogive (percent):

Age (year)	1	2	3	4	5	6	7+
% $matures_i$	34	90	100	100	100	100	100

The estimated recruitments and the spawning biomasses between 1986 and 1996 are presented in the following table :

Table 1

Year	N (age 1) (thousands)	Spawning biomass (t)
1986	8751	1957
1987	8305	1591
1988	7123	1956
1989	7596	2073
1990	6013	2287
1991	5054	1506
1992	9713	1400
1993	5520	1275
1994	6000	980
1995	7329	675
1996	6840	917

1. Calculate the biological reference point $F_{0.1}$. Indicate the value of the virgin biomass that you estimated and calculate the percentage$B_{0.1}/B_{virgin}$..
2. Calculate the biological reference point F_{max}.
3. Calculate the biological reference point F_{med}
4. Estimate the parameters of the S-R Ricker model and indicate what the value of the spawning biomass/recruit is, corresponding to F_{crash}.
5. Comment on the present state of the stock and its exploitation.

QUESTION 2

Consider a stock where the length compositions of the catches during 1986-1995 are known (Table 2).

The following parameters were estimated from this stock :

Natural mortality coefficient = 1.2 $year^{-1}$

Asymptotic length = 39.8 cm

Growth coefficient= 0.8 $year^{-1}$

Table 2 – Catch Matrix (thousand individuals)

Length classes (cm)	Year									
	1986	**1987**	**1988**	**1989**	**1990**	**1991**	**1992**	**1993**	**1994**	**1995**
14-	10	9	8	9	7	18	13	12	15	8
15-	20	17	15	18	13	35	26	23	29	16
16-	29	25	23	26	19	52	39	34	42	24
17-	47	41	37	43	31	84	63	56	69	39
18-	92	80	72	83	60	164	122	108	134	76
19-	224	194	175	202	146	398	297	263	325	185
20-	261	226	203	234	169	461	343	305	376	214
21-	420	363	326	376	271	736	547	488	601	342
22-	335	525	603	506	511	365	983	571	657	741
23-	345	540	618	516	520	370	995	581	666	752
24-	422	661	751	622	625	444	1189	699	799	902
25-	442	693	781	642	643	454	1212	718	818	923
26-	415	650	726	592	590	415	1102	659	747	843
27-	388	607	672	542	537	376	995	601	677	765
28-	360	564	617	493	486	339	890	543	609	688
29-	332	520	563	444	435	301	788	487	543	613
30-	304	475	508	396	386	265	689	431	478	539
31-	275	430	453	348	337	230	594	377	414	468
32-	272	246	509	494	330	314	212	428	351	353
33-	239	216	439	419	277	261	174	359	292	294
34-	206	186	369	345	226	211	139	293	235	237
35-	171	155	300	273	176	162	106	228	181	182
36-	136	123	230	202	129	117	75	167	130	131
37-	82	99	116	196	147	91	80	40	117	83
38-	49	60	66	103	75	45	38	21	58	41
39-	0	0	0	0	0	0	0	0	0	0

1. Estimate the values of the fishing mortality coefficient in each length class of the cohort of 1987 (to simplify, adopt $F_{terminal} = 0.5$ year^{-1}).
2. Estimate the corresponding values of the number of survivors at the beginning of each class.
3. Say what the recruitment of this cohort is.

QUESTION 3

The Fox Production model was adjusted to a certain stock and the following parameters were obtained :

$K = 300$ thousand tons
$r = 0.50$ year^{-1}

1. Determine the biological reference points F_{MSY} and $F_{0.1}$.
2. Knowing that in recent years the biomass of this stock is about 30 percent of the virgin biomass, comment on the present state of the stock and its exploitation, based on the adopted model.